The Kingdom *of* Infinite Space

The Kingdom of Infinite Space

Infinite Space

A Fantastical Journey Around Your Head

Raymond Tallis

ATLANTIC BOOKS
LONDON

First published in hardback in Great Britain in 2008 by Atlantic Books, an imprint of Grove Atlantic Ltd.

The author and publisher would gratefully like to acknowledge the following for permission to quote from copyrighted material:

'Ignorance' in *The Whitsun Weddings* by Philip Larkin, Philip Larkin © 1964 reproduced by permission of Faber & Faber for the author; *The Art of the Novel* by Milan Kundera, Milan Kundera © 1988 reproduced by permission of Faber & Faber for the author; 'The Love Song of J. Alfred Prufrock' in *Collected Poems 1909-1962* by T.S. Eliot, T.S. Eliot © 1963 reproduced by permission of Faber & Faber for the author; 'Dante Alighieri' in *The Penguin Book of Italian Verse* edited by George R. Kay, George R. Kay © 1958 reproduced by permission of Penguin; 'First Night' in *Nights In The Iron Hotel* by Michael Hofmann, Michael Hofmann © 1983 reproduced by permission of Faber & Faber for the author; 'Little Gidding' in *Collected Poems 1909-1962* by T.S. Eliot, T.S. Eliot © 1963 reproduced by permission of Faber & Faber for the author; 'Tonight at Seven-Thirty', section X in 'Thanksgiving for a Habitat', in *Selected Poems* by W. H. Auden, W. H. Auden © 1968 reproduced by permission of Faber & Faber for the author; 'I Am Not a Camera', in *Selected Poems* by W. H. Auden, W. H. Auden © 1971 reproduced by permission of Faber & Faber for the author; 'Skin' in *Collected Poems 1909-1962* by Philip Larkin, Philip Larkin © 1988 reproduced by permission of Faber & Faber for the author.

9 8 7 6 5 4 3 2 1

A CIP catalogue record for this book is available from the British Library.

ISBN: 978 1 84354 669 6

Designed and typeset in Monotype Dante by Lindsay Nash
Printed in Great Britain by MPG Books Ltd, Bodmin, Cornwall

Atlantic Books
An imprint of Grove Atlantic Ltd.
Ormond House
26–27 Boswell Street
London WC1N 3JZ

www.groveatlantic.co.uk

To Ben
Jedné z mých nejdražších hlav

Sustained in the body given him by nature between those two abysses of the infinite and the nothing.

Pascal, Pensées

Contents

Acknowledgements

I am enormously grateful for the support and enthusiasm of Toby Mundy at Atlantic without whom I would not have written this book: it is his idea. It has been wonderful to work again with Louisa Joyner, whose brilliant editorial skills have transformed this book, not the least by giving the author a better understanding of what he was trying to achieve through it. Latterly she was joined by Emma Grove who had a crucial role in bringing the manuscript to its final form. Finally, it is a pleasure to acknowledge the superb copy-editing by Jane Robertson. What luck to have been published again by Atlantic and again to work with such a team.

My greatest debt is, as always, to Terry, my wife.

Foreword

The Kingdom of Infinite Space owes its existence to Toby Mundy of Atlantic Books. He had suggested that I should write a book on the body, which would encompass biology and philosophy. The more I thought about it, the less manageable the topic seemed and the greater the danger that philosophy would be buried under biology. I was scratching my head as to how to deliver on his suggestion when it occurred to me that the answer lay under my fingernails. I would confine myself to the head.

Why the head?

Well, of all the items in the world, my head is the one that seems closest to me, in the rather difficult sense of being what I am, or (not quite the same thing) what I feel myself to be. And yet my relationship with my head is not at all straightforward. I am linked in different ways to different parts of it; and I have different links at different times to the same parts. Yes, I *am* my head (in a sense that proves extraordinarily difficult to characterize). But I also own it, speaking of it as I do now as if it were a possession. And I suffer it, enduring aches and pains that seem to have nothing and everything to do with me. I use it, manipulating it, sometimes quite crudely, as if it were a kind of tool. I present it. I judge it. I know it. I disown it. And so on.

All very baffling. 'My head and I' is a more problematic marriage than anything that August Strindberg or Edward Albee could have dreamed up. In this regard, however, my relationship to my head is no different from the muddled, even tortured, relationship I have to my body as a whole; indeed, it is the same relationship put in capitals. Which is I why I have chosen to write about the head: it is an entrée into the peculiar human condition of being a conscious and self-conscious, self-judging, agent in an organic body. I want to think into the muddle of embodiment. And I want to celebrate the mystery of the fact that we are embodied, rather than fall in with the venerable tradition of being rather sniffy about it.

Many writers, following the example of Plato, have seen the body as a kind of prison, a cognitive disaster, a humiliation, or some kind of moral disgrace. I, on the other hand, while I do not believe that we are immortal souls, unhappy lodgers accidentally trapped in 70 kilograms of protoplasm, equally reject the notion that we are entirely identified with our bodies. The standpoint of this book is neither religious nor scientistic, but humanistic.

Defending humankind against supernatural and naturalistic accounts of itself is a preoccupation informing *The Kingdom of Infinite Space: A Fantastical Journey Around Your Head*. This preoccupation will not, however, always be present or explicitly so. Because there is a lot that is wonderful and funny about my head, I see no reason to resist a lifelong habit of digression. (The poetry of existence, Kundera said, summarizing Laurence Sterne's philosophy, lies in digressions.) If the overall impression is of an album of sketches, then I won't be entirely unhappy. After all, the philosopher Ludwig Wittgenstein described his posthumously published masterpiece *Philosophical Investigations* as 'an album' and as 'sketches of landscapes made during the course of... long and involved journeyings'.[1]

The Kingdom of Infinite Space, too, endeavours to do philosophy of a sort. Wittgenstein once described his philosophical method as being to 'assemble a set of reminders for a purpose'. In his case, the purpose was to remind people of what was in front of their noses, so that they would not be bewitched by language into engaging with insoluble pseudo-problems. This did not deliver what he hoped. And a good thing, too. Some of the things he considered to be pseudo-problems, ripe to be 'dissolved' (rather than solved) by appropriate linguistic analysis, are handles by which we may take hold of, and interrogate, the too smooth surface of our lives and consciousness. So, although this book, too, is often an assembly of reminders, its purpose is to heighten, rather than allay, astonishment and the sense of mystery.

In some places its approach may be best described as a rather informal phenomenology – 'phunomenology', even – reminding us of what is behind as well as in front of our noses. Phenomenology focuses first and foremost on the 'phenomena' – a word derived from the Greek, meaning 'appearings' or 'appearances' – and is motivated by the sense that our immediate experiences are the proper object of philosophical attention. Such experiences are 'the given'; whereas factual knowledge, in particular scientific knowledge, is derivative, and consequently questionable, rather than the final authority. I am very attracted by the idea of phenomenology as a philosophical method and ambition, even though things have not worked out as the great phenomenologists – Brentano, Husserl, Heidegger, and Sartre – had hoped. They built huge castles of rather abstruse prose which seem somewhat remote from the immediate experiences they are about. What's more they have proved unable either to account for, or to take account of, the mighty truths of science. What is the relationship between what I am feeling now and the scientific account of the world that has had such an impact on what I am feeling now?

You won't find an answer in the many fascinating pages of the phenomenologists. Edmund Husserl admitted as much towards the end of his life. He could not find a way of linking the natural science that aims to give an objective account of what is out there with the subjective flow of individual consciousness. It wasn't for the want of trying. Many long, densely written books and 45,000 pages of unpublished writing testified to this and to the tragedy of his cry, three years before he died a heartbroken technician and visionary:

Philosophy as science... *the dream is over.*[2]

This failure touches on something central to this book: the dissociation between facts and experiences. *The Kingdom of Infinite Space* is haunted, or at least bothered, by this: our experiences of our head are not fact-shaped. We cannot bridge the gap between what we feel ourselves to be, how we experience ourselves from moment to moment, and the innumerable facts, ordinary, extraordinary, and recherché, about us. Many, perhaps most, of the facts in this book go a long way beyond the immediate data of experience.

There is a lovely story about Jean-Paul Sartre and Raymond Aron, meeting in the Café de Flores. They were at the outset of their dazzling careers and they had not yet had their bitter falling out. Aron had just returned from Germany where he had sat at Husserl's feet and he brought news of the phenomenological movement back to Paris. He pointed to a glass on the table and said to Sartre: 'That is all you need in order to do philosophy.' No equipment or technical know-how or esoteric knowledge was required; just an ability to reflect on the experiences one was having. One's subjectivity is the appropriate point of departure for philosophy. Sartre is said to have gone pale with excitement. I continue to share that excitement, if not the pallor.

Looked at in a certain way, the head provides a wonderful entrée into many of the more traditional philosophical themes whose scope extends way beyond that marvellous structure. The reader may therefore look forward to what I hope will be user-friendly glances at topics such as the unique freedom of human beings, self-knowledge, the nature of personal identity, vagueness, and much more besides. We will head off along many paths.

One path I shall generally try to avoid is the one that ends in 'neurophilosophy' – the currently dominant theory that says that the mind or consciousness is identical with the neural activity of the brain. This book about the head says little about the brain. I imagine some readers will be glad to learn this and be more, rather than less, inclined to read on. There may, however, be one or two who feel that a book about the head that does not give the brain a starring role is *Hamlet* in which the Prince has only a walk-on part. After all, the brain is not only the biggest item in the head but also the most talked-about. It is to these readers that the following comments are addressed.

First, be assured the importance of the brain has not entirely escaped my attention. I know that if I mislaid mine, my IQ would fall quite disastrously. Indeed my living daylights, both intelligent and dim, would go out and their support services grind to a halt. The fact that head injury has a more profound effect on what I am than leg injury has something to do with the fact that my head contains my brain. There is, however, no shortage of books on the brain. Indeed, I would venture that there is a serious lack of such a shortage.

Try searching the net as I have done just now. The result is a Niagara of logorrhoea: 'Brain and Consciousness' 541,698 items; 'Brain and Self' 2,114,747 items; and 'Brain and Mind' 2,939,316. Amazon offered me 837 books dealing with brain, mind, self and consciousness and 60,970 items somehow implicated with the brain. If you key in 'Brain' alone you will find 11,351,398 items on the net, all eagerly awaiting your hit.

As the philosopher J.L. Austin once said, one has to be a special kind of fool to rush in where so many angels have trodden already. But a saturated market for books on the brain is not the only reason why most of the present volume will be resolutely extra-cerebral. To put it bluntly, the brain is absurdly overrated. The reason why there are so many books being published, and even read, and certainly being remaindered, on the Brain and Consciousness, the Brain and the Mind, the Brain and the Self, the Brain and You, is that there is a myth out there that the explanation of consciousness, the mind, the self, and you is to be found in the brain.

Some people imagine this is a new idea. It was old even when Hippocrates (500 BC) asserted that

> Men ought to know that from the brain, and from the brain only, arise our pleasures, joys, laughter and jests, as well as our sorrows, pains, grief and tears. Through it in particular, we think, see, hear, distinguish the ugly from the beautiful, the bad from the good, and the pleasant from the unpleasant.[3]

Here is not the place for a detailed account of why this Greek myth – the central myth of what I have called 'neuromythology' – is wrong. The bottom line is this: the brain is a *necessary* condition of all forms of consciousness, from the slightest twinge of sensation to the most exquisitely constructed sense of self. It is not, however, a *sufficient* condition. Selves are not cooked up, or stored, in brains or (as writers such as the late Francis Crick would have it) in bits of our brains, such as the claustrum. Selves require bodies as well as brains, material environments as well as bodies, and societies as well as material environments. That is why, despite the hype, we won't find in the brain an explanation of ourselves, or the secret of a better self or a happier life.

I have devoted several books and quite a few articles to setting out why consciousness is not to be found in the brain, which the interested reader may wish to consult.[4] And I have argued for a period of silence while those scientists and philosophers, and others smitten by the brain, digest these rather obvious facts and think about how the gap between the necessary and the sufficient conditions of consciousness, mind and the self is to be crossed and with what.[5] For the most part, this book observes that period of silence. However, out of fairness to the reader – and to the brain, and to the neurotheologists who worship it – I do visit various aspects of the mind-brain identity theory at relevant places in this book. I have always felt that philosophical arguments make more sense if they are located in an appropriate context. Indeed, I would almost go so far as to claim that, diffused through this book, is a comprehensive critique of brain-centred understanding of human beings and their consciousness.

At any rate, if this book is about a largely brainless head, it is a corrective to the headless brains that fill the newspapers, bookshelves and the airwaves. In essence, *The Kingdom of Infinite Space* is an attempt 'to get my head round' the head.

My coverage of the subject will be highly selective. I will focus on a relatively small number of topics, somewhat artificially gathered under a handful of headings (the term proves inescapable). I have quite a lot to say about the uses to which breathing is put and less to say about eating. Secondly, the coverage of each of my chosen topics will be far from comprehensive. While there is quite a lot of biology, *The Kingdom of Infinite Space* is no more a textbook of cephalology than Primo Levi's *The Periodic Table* is a treatise on chemistry. The literature on speech, on facial expressions, on hair, even on earwax and head-butting, on all the topics I touch upon, is boundless and, with respect to it, I have only the Socratic wisdom of knowing how ignorant I am.

But the head is far too interesting to ignore the extraordinary facts established by scientific inquiry. Anyone who wants (as I do) to be serious about saliva cannot avoid discussing either semi-permeable membranes or the history of spitting. This book will again and again lose its way, which happens on all journeys that are worth undertaking.

My failure to do full empirical justice to the head is, however, consistent with the aim of this book. There is no end to the things that I could talk about and no end to what I could say about them, but I am mindful of what Goethe said somewhere: we should not know more of things than we can creatively live up to. There is a gap between the near-infinity of facts about the head and the limits to what we can make imaginative use of. Dumping everything that is in my head, or the collective head, into yours via a book is no way to proceed.

My choices will be guided by an informing purpose: to make the head more visible to its owner, in part by highlighting the rather odd relationship we have with our own heads. I hope that the sheer oddity of our headedness will become apparent in the chapters to come. If my readers end this book as astounded tourists of the bit of the world that is closest to being what they themselves are, I shall be content. All you have to bring to the party is your own bonce, in reasonable thinking order.

Let us now head off into the known that it may be known better – or at least differently.

chapter one
Facing Up to the Head

Could any man bear to look at himself at every
moment of his life, and rethink, as a witness, all he has
thought, all that has come into his head, into his whole
being? Who would not hate himself, not wish to blot
out what he was, not so much from want of success,
or the effect of certain acts he has committed, but
simply because of the *particular person* whom these
have little by little *defined*, and who shocks his full
sense of possibility. Our history makes of us Mr *So-
and-So* – and this is an offence.

Paul Valéry, *Monsieur Teste*

Portrait in an Ordinary Mirror
Look into a mirror. Nothing could be more routine. It is something
we do without thinking, every morning, when we prepare our
faces for the judgement of the great world. And yet it is an extraordinary thing to do. Nothing could be less straightforward than the
relation between our head and its mirror image.

You are not René Magritte, so it will be the front of your head,
your face, that you will see mainly. You may note that your head –
Everyhead – is of modest size, neither egghead nor kingly nor
microcephalic, justifying neither a second thought nor a further
adjective. You may be generally pleased with it, displeased, or indifferent. Most likely, it will be one of those middle-of-the-road heads,

with a face that will neither make your fortune nor count as a great misfortune. One of those faces which strangers neither look at in longing or in horror, or in fear. In short, for philosophical purposes, an ideal face: the face of Everyman or Everywoman.

Perhaps you might try to describe yourself, penning a verbal portrait of the portrait in the mirror. This is what I see, looking out at me. The longish, sunken-eyed face is that of a man in his late fifties. And, frankly, it looks it. The time-weeded cranial dome is encircled by a low hedge of grey hair, echoed in the grey Van Dyke beard. Above the eyes are those frontal corrugations some of which, echoing the curves of the eyebrows, mock both the brows and half a lifetime of surprises that, by dint of repetition, have been translated into fixed structures. The eyes have mud-yellow sclera and irises of an indeterminate greenish bluish brown (I have had to check in a better lit mirror). There is a slight drooping of the eyelids, as if the effort of keeping my eyes open for so many years is starting to tell.

The sitter for this instantaneous self-portrait has a straight nose, from the side a perfect half isosceles, marred only by a slight asymmetry of the cartilaginous tip. The large ears look as if clamped on as an afterthought. The mouth is not generous and the less these lips say of themselves the better. Through them, he pokes a pink tongue – his palate's closest pal – in order to remind himself of his existence, and then retracts it in order to expose his teeth (his own and in reasonable condition), all encircled by a full-set beard.

You try it now, putting your own face in a hole cut in the page. You will at once realize, as I have, that we are facing defeat. The defeat to which I refer does not lie in the face itself (though there is a little of that as well) but in the sure-fire failure of any attempt to transfer the face from the mirror to the page. The face that looks back at the gazing face – with that very gaze, into that very face (an easy but misleading model of self-consciousness) – lies beyond any descriptive powers. Your face is a singular entity and your words are general. Any verbal portrait you might construct will be little better than those Identikit pictures that the police issue when they are looking for someone with whom they would like to have a conversation. The transformation of a visible surface into an intelligible description is no easier even where the surface in question is your own face. The more it makes general sense, the less faithful it will be to its individual appearance.

Writers often give us the impression that they have described the faces of their characters, when in fact they have simply given you an outline to fill in. Of Esch, the most important character in Hermann Broch's masterpiece *The Sleepwalkers*, we learn only that he has big teeth.[1] Even so, we don't feel as if his face is a dentate blankness. Most often, we mistake being told what effect someone's appearance has for an account of that appearance. The poet Mallarmé's advice – *Peindre non la chose, mais l'effet qu'elle produit* ('Paint not the thing itself but the effect it produces') sounds like a self-denying ordinance.[2] Actually it is a rather cunning way out of an intractable problem. When, in one of his novels, Evelyn Waugh says of a new character, that 'he had just the kind of appearance one would expect a young man of his type to have' and nothing else, you still feel as if you have been told exactly what he looks like.[3]

If you have stared at your face long enough to feel that it eludes all the words that would locate it snugly in the world, you may start feeling odd. The fact that this face is my face, this head is *my* head, will dawn on you. Fancy, you think, of all the faces I know, being this face. Fancy, come to that, being *this* thing, living *this* life. The feeling passes: after all this head *is* your head; the thing in the world that is nearest to being what you are. Nothing could be more familiar or an object of more anxious concern. And yet you know so little of it.

Look at your head and remind yourself of its multiplicity: its many components and the multiple uses to which some of them are put. The head is a site of endless trafficking. It takes in and it gives out 24/7. There are inputs of sense experience, of air, and of food: it harvests sights and sounds and smells and tastes; sequesters indoor and outdoor, private and public, urban and rural air; and ingests food and drink and medicines and worse. And its outputs are as impressively varied. The number and variety of secretions should compel admiration, though our attitude to them is some-what ambivalent. They range from fluent ones such as saliva and tears, to more measured examples such as ear wax, to the glacier-slow growth of hair and of teeth. There are other outputs which leap straight out of the head – vomiting, for example. More impor-tantly, the head emits an endless variety of signals, voluntary and involuntary, linguistic, paralinguistic (such as affirmative nodding) and non-linguistic (like smiling). All of these will detain us in due course. However, before we set out on our journey around the head, let us think what purpose such a journey might serve.

Dwelling on our relationships with our heads is a way of getting hold of our relationships with ourselves: what it is to *be* this self. This relationship is highlighted when we look in the mirror. The first, and most obvious, thing to be said is that the gaze that is looking out at your head is also the gaze that is looking in at your reflection. It is your gaze and it is intersleeved with itself, in a chaste, ocular auto-copulation. This seems promising, a perfect philosophical thought, though difficult to maintain. The mental gaze – unable to stand still, unlike the long-legged fly on a stream or a kestrel at stoop – wanders. Other things, however, may strike you.

For example, the head you are looking at when you gaze in the mirror is silent and yet you can hear your thoughts articulating themselves and you are inclined to locate them 'in my head'. (Big trouble for you if you don't: mislocating or misallocating your thoughts brings the gentlemen in white coats, armed with syringes.) You could not, however, tell from looking at your head that it *is* thinking, never mind *what* it is thinking.

The stared-at head is as opaque to its owner as it is to other people or others' heads are to it. In fact, as we look at that opaque object elevated by the neck above the shoulders, it suddenly becomes difficult to tell that it is looking, even less what it is looking at, less still what it succeeds in seeing. (As philosophers remind us, with their charming pedantry, 'seeing' is not a state but 'an achievement'.) It is only because it is *your* head that you have no difficulty seeing that it is seeing, or being confident of your thought that it is thinking. Indeed, you cannot doubt this, as Descartes argued, since to doubt that you are thinking is to think. Actually, Descartes' 'I think therefore I am' does not take us very far. The logical certainty you have that you are thinking does not guarantee that the head that is thinking is the one that is looking out at you from the mirror and which smiles when you feel yourself to be smiling.

This prompts a question. How can Everyhead staring in the mirror tell that this head in the mirror is *its own head*, in the way that it feels it to be its own? No answer is forthcoming. Nevertheless, the owner feels entirely easy and at home with identifying his head in this fashion and calling it 'my head'. Heads may shock, disappoint, please, and worry their owners, but they never ambush them with the fundamental question: Whose head? This ownership goes very deep indeed – to the very bottom of what we are.

OK. But what does it mean to say that this head staring out of the mirror belongs to the person staring back at it? Or, to vary the question, what does it mean to say that you *are* this head? At the very least it means that you suffer, or experience, this head without mediation. It is the site of the sensations you have now: the down of dry warmth on your cheeks; the weight of the upper teeth on the lower, falling just short of a clench; the seeping of saliva behind the teeth and under the tongue; the verge-of-headache dazzledness around the eyes; the looking through the nose-supported spectacles and the pressure of said spectacles on said nose; and so on. This is how you experience the very special relationship between that object with the visual appearance you see in the mirror and the experiences just listed: they refer to – they are of – the same thing.

We could frame the question differently, perhaps at a more superficial level. How do you know that the visual impression of the face that you see in the mirror is of the same thing as certain sensations you may be feeling – for example the diffuse headache that seems to have developed in response to your attempts to take up my invitation to concentrate on what it is to be or to have this head as your head? This is something that has preoccupied philosophers.[4] The great German philosopher Immanuel Kant was especially puzzled. How is it, he asked, that different sensations, arising from the different senses and seemingly originating from different parts of the body, and occurring over time, are felt to belong to the same moment of consciousness of the same person? What holds them together? What makes them all belong to the same moment of the same person? And what makes successive time-slices of persons belong to the same person, so that there is a smooth transition between our moments? How is it that at any given time, 'I' am enjoying a tune, feeling the weight of my bottom on a chair and seeing the blackbird flying past the window? And how is each moment of the tune impregnated with other moments, so that I can enjoy the successive notes as parts of a melody? Kant called the necessary binding 'the unity of apperception'. Being somewhat cerebral, he spoke of the '"I think" that accompanies all my perceptions'. Vision, touch, smell, all the buzzing confusion of the moment of consciousness, were referred to the same me, he said, because each of them was accompanied by a reference to an 'I' that thinks it is the same 'I'. I am not totally convinced that that is quite right: the tactile down on the cheek does not seem like an endlessly repeated 'I think'. Consciousness is not so donnish. Nor does 'I think' capture what it is that binds that faint sensation on the cheeks to the gaze of our head at itself in the mirror, or to awareness of the thoughts we are having.

As I gaze into the head that is gazing at me, into the thoughtful face of the man who is thinking these thoughts, which include the thought that the face is thoughtful and that it is the face of the man who is thinking the thoughts, vertigo beckons. And while philosophy is, quite properly, a dance around the edges of a whirlpool, it is probably a good idea, once you feel the current taking hold, to take precautions against drowning.

Avoiding Anatomy

Let us cover up the mirror and return to the physical reality of the head, unhollowed by introspection. There is a massive body of knowledge about the head, spread over thousands, perhaps hundreds of thousands, of textbooks, websites and databases – and heads. Most of it will apply to all the 6,000 million heads on the planet, the 6,000 million objects from whose open mouths pops out the first-person pronoun. But we mustn't be over-impressed by facts, for several reasons.

First, very little of our experience of our heads is fact-shaped. The myriad of fugitive sensations by which my head declares its presence elude being skewered by sentences. Secondly, there are many facts about my head that are not experienced by me. I defy anyone to feel the measurable difference in manganese levels between tears prompted by a poke in the eyes and tears provoked by grief. Thirdly, there are even more facts that are neither experienced by nor known to me.

It is astonishing to think how little of our head is available to us other than by third-person report. This is true not only of recherché matters, such as the microscopic structure of the vasculature in the skull vault or the way the bones in the middle ear are connected up. It is also generally true of pretty easily available stuff. Most of my skull vault is at best faintly present. Facial and cranial skin glows only intermittently and on a rather low flame of awareness. Ears tend to come out largely in winter, when the rim of the pinna is nibbled by the piranha fish of cold air. My brain, the biggest thing in the head, is silent for most of the time. When it does speak, its locutions tend to be referred elsewhere: activity in the brain is 'about' the non-cerebral body or the world. In short, the presence of the head to its owner is an intermittent, spatially discontinuous blossoming out of absence.

This usually suits us just fine. We can manage nicely with a head that has disappeared largely down a hole in consciousness. The head seems to work best as a justified assumption, which is cashed only intermittently, rather than an asserted presence. The general silence of the skull vault, for example, does not undermine our confidence that, when we attempt to head a football, it will bounce towards the goalmouth rather than fall straight through the brain, into the mouth or whichever part of the head is currently on parade. And when we open our mouths to eat, we can assume without checking that the palate of which we are currently unaware will be there to guide the food to the right place.

If all the places in the head that might be called upon to do their duty were required to be constantly iterating their presence, there would be such a cacophony of cephalic sensation that it would be difficult to see how the necessary attention could be paid to the parts that mattered or any attention could be paid to anything that was not a part of the head. If the tongue was constantly aware of itself, language would be drowned in the auto-Babel or babble of this busy piece of meat endlessly discoursing of itself. The background silence of the skull vault is necessary to ensure that the contact with, or the failure to contact, the football would be registered.

Of course, there are times when places that should be quiet are noisy. The accidentally bitten tongue shouts out horrible anti-meaning. When the playground bully tugs at the earlobe, he gatecrashes our sense of what we are – that is why it is such a potent means of humiliation. The throbbing vertex makes the scalp present when it has no job to do. The aching tooth develops an agenda at right angles to the purposes, the tasks, the hours of its owner. Tinnitus, courtesy of which the head is a source of noise rather than a device for detecting it, curdles the heard world. These are potent reminders of the importance of the silence of the head.

Some of this silence is absolute and some is merely relative, a case of being overshadowed by more clamant parts. That is why we can awaken less assertive parts by an inward shift of attention. Consider how it is when you are absorbed in some activity – say peeling sprouts. You are so busy with your hands, your mouth goes unsensed. Now deliberately wake up out of the task and attend to your oral cavity: you become explicitly aware of what may have been quietly present but ignored. You feel: the little pool forming under your tongue; the way the upper surface of your resting tongue conforms to your hard palate, soft curve to hard one; a very faint throbbing of the dentate horseshoe of your lower gums; the pressure of the lower teeth against the upper set resting idly on them, and so on. And just to the north-east of your mouth, an intermittent glistening of air through your nostrils as you inspire.

Redirecting and focusing attention can take you only so far. No amount of concentration will wake up our hair unless it is dis-placed. Scalp hair will require a breeze for it to be registered as a presence on the skull – and then it is the tugged-at skull rather than the hair that does the registering. Of course many structures are woken up when they are engaged in their legitimate activities – otherwise we would have no way of regulating and directing those activities. Whilst it is mostly unconscious, feedback has to be conscious for us to know what we are doing, how far we have progressed, and whether we have done it. So dormant patches of our mouth awaken when we drink a cold glass of orange. We can even track it down to places that the most concentrated attention could not otherwise reach, well past the point at which the throat divides into windpipe and gullet. It is a torch, momentarily lighting up the darkness within the body.

All of this has profound consequences for our sense of what we are and hence for what we are. To see this darkness at your heart, it is necessary only to close your eyes.

Try it.

Ego-Head: Elementary

In order to *be* something, at the very least we have to experience it. We have no or intermittent experience of the vast bulk of our heads, from which it follows that we are *not* most of our heads. Nevertheless, the head seems to be the capital of the first-person world. It lies at the centre of its inmost circle. The self – the ego – seems to be closer to the head than to any other object in the world. So how is this to be explained? Perhaps when we say that we are in our heads, we mean that we are the gross and fluctuating outline but not the details. We are not, at any rate, everything that is revealed to us about our heads by observation, in particular the special observations of the scientist. We are not our individual blood cells or our serum potassium. In short, 'I' is not constructed from the facts of its case. Egos are not things that occupy space while heads do occupy space; ergo egos cannot be located in heads. Next question please.

Not so fast. If you were to ask me where I was, I would say I was here. And if you asked me where 'here' was, I would say that it was where my body was. That, at least, lies at the centre of a set of concentric spheres into which my awareness – my sensory field, the field of my knowledge – extends. Inside Cheshire, there is Bramhall; inside Bramhall there is such and such a road; within this road, there is house number 5; inside number 5, there is my study; and inside my study there is my body, at my desk. This is where I am. I am where my body is, though my sense of where that body is (in 'Cheshire', for example) is projected on to a very complex network of conceptualized places. If you were to ask me where *within* my body, I was, my head would seem to be a more promising candidate than, say, my spleen or even my leg.

There is a reason for this of course. *Pace* the neuromythologists, it is not because I am my brain and my head is where my brain is. It has much to do with the special involvement of the head with the perception of objects outside of the body. Many other parts of the body are endowed with proprioceptors – senses that report the body's experience of itself and of its movements, its position, its location and its actions. And most other parts of the body have surfaces that are equipped with tactile senses – light touch, heavy touch, painful touch, warmth, cold, tingling and so on. And while the hand is the chief organ of the fifth sense, it has no monopoly of touch. Heads, too, touch, as kissers (*vide infra*) know only too well. The head, however, has a monopoly over sight, hearing, smell and taste. Heads see, hear, smell and taste the way that legs do not. What we see, hear, smell and taste depends on the state of my head and, more relevant to our present theme, its location. Seeing and hearing (and possibly smell, but this is less important for us humans) are telereceptors: they uncover objects at a distance. Vision is the supreme telereceptor: its objects are located in a place with respect to other objects.

What has this got to do with the head and the location of the ego? Very simply this: the objects that are located over there stand in relation to you. (Except when the objects in question are conscious human bodies, they are not aware of this relation, of course.) You locate them over there and they locate you over here. The objects that you see arranged around you in a visual sphere situate you at the centre of that sphere. You (as 'I') are the reference point that classifies things as 'near' and 'far'; as being more here than there or more there than here; as handy or out of reach; as in the frame or off the map. Your body is that around which everything is arranged. And the part of the body that lies at the centre of this centre is your head because this is the refer- ence point of the visual field: this is where the rays of light that come from the objects that you see converge. This is where they are harvested. This is where they are delivered. This is their destination.

When you see an object, you not only see it but you see it in rela- tion to other objects; and in relation to your body which is itself an object, though it is of course much more than that. You also see *that* you see it. Your seeing body thus presents itself as your point of view on the world. And it is obvious that, within your body, it must be the part that does the viewing – your head where your eyes are located rather than your feet – that is the viewpoint. Thus is the head located at the centre of what that great man, Professor Edmund Husserl, called 'egocentric space'.

Now a word of caution and a couple of observations before we proceed with trying to understand how you and your head get so close to being identified with one another. 'Egocentric' space must not be confused with physical space, the space of physics and math- ematics. First of all, the visual field is not merely an array of objects set out in geometrical locations. It is also a network of sig- nificances; of things that may or may not be of interest to the individual. As Martin Heidegger (Husserl's greatest pupil) pointed out, things exist for us in a world which in part we constitute, as things ready-to-hand rather than as mere physical objects that are present-at-hand. As you are busy preparing a meal, the space in which you are located is not a set of objective distances linking objects of equal standing. It is subsumed into a cosy parish of handiness – of things waiting to be reached for, to be knocked over, to get in the way, or simply ignored, being out of the range of your attention or interest. The handiness of something is not merely a matter of its being within a physical distance defined by the length of your arm. You will experience something as handy if it is not only within reach but also of some interest to you.

The consequence of this is that the egocentric space of which we are the centre is not narrowly defined by a visual space that can be understood in physical terms. After all, the rest of your body – the part of the body that reaches out, pushes and shoves, and walks towards and away from – is also implicated in establishing egocentric space and in installing you as the centre of that space. And so, by the way, is some of the rest of your life, which sees the coffee as unfinished and the biscuits crumbs as disorderly and the letters on the computer screen as the object of your solicitous attention. The roll call of objects that are related to the centre fluctuates wildly; and the extent to which the ego dissolves into that wobbly space, and the parts into which it dissolves, also vary. The sense of the I is lost in, and wakes out of, the world with which it engages.

This goes some way to mitigating the sense that the centre of the space – the head – is a somewhat eccentric centre. The solar plexus, or some such midpoint in the body, would be literally more appropriate. This, of course, would not be a very good place to locate the eyes. It is because they are near the top of the body, hoisted up on five or six feet of meat, that their viewpoint becomes a vantage point, and the body they serve becomes a watchtower.

This does underline something else: that the centre of ego-centric space is not a sharp point. It is blurred and smeared; it is spread to a greater or lesser extent over the body. The fact that we have two eyes and two ears reinforces the head as the centre of the centre of the meaningful space that surrounds us. If it does not result in a further smearing between two eyes that are themselves not punctate and two ears that are separated by the width of the head, it is because the directions from which that which is seen and heard seem to converge on an implicit vanishing point. After all, the experiences that make our experiences of our bodies all add up to the moment of 'I' consciousness. There is, as we noted before, in Kant's words, a unity of apperception. We may think of this as the 'convergence point' of our attention. And it is this that is the location of the attending self – the centre of egocentric space.

It is worth thinking about the unargued intuitions that lie behind this notion of the 'convergence point'. There is the idea that all the things that attract our attention are, as it were, pointing themselves out to us. They draw us in different directions – this is the etymo-logical basis of 'ad tend' – to draw towards. We then lie at the point of origin of all the lines that connect us with the objects of our attention. The aware body transforms that which merely is into something that is 'here' or 'there'. The head is that part of the body which does most to define the here, because it is the locus of those telereceptors – of sight, primarily, but also sound – that disclose what is as that which is there. The ego is the 'here', the 'here of heres', 'the heremost here' among those things that are 'there' (for me).[5]

And so, when I reflect on my self, in that artificial state when we do philosophy, and try helplessly to catch ourselves unawares, we locate ourselves just behind our eyes and perhaps above our mouths, in a little virtual space from which we taste the world. This is where we are, we think. But is it where we think? Answering that question opens up more fascinating questions which we shall reserve until later, for fear of losing ourselves too early in abstractions.

It is time now to peer inside the head.

chapter two

The Secreting Head

Stuffs Happen

We have direct experience of only a limited part of our heads – our awareness penetrates only so far – and this seems to be a barrier to our identifying ourselves with them. There is another barrier almost as great: our heads transact an enormous amount of business which, though it is to our benefit, is performed without consulting us. The head has itself, and an attached body, to look after. Actually, the relationship is reciprocal: the head looks after the body so that the body will look after the head. The essential point, however, is captured in the poet Philip Larkin's rueful observation that 'our flesh/ Surrounds us with its own decisions';[1] this is as true of the flesh that is our heads as of the flesh that is our bottoms or our spleens.

Nothing could highlight this more dramatically than the way our head expresses itself in a variety of secretions that it, not we, secretes. Which is a good thing: if we had to *do* the secreting that happens in the head, we would be in serious trouble. For the secretions are not optional extras. It is fortunate, therefore, that mechanisms such as those that underpin secretions are at work in the very headquarters of the human agent, where free action is forged; that stuff happens right where we act.

The range of stuffs that happen in and around our heads is impressive. There are the obvious ones such as saliva, sweat and tears; and the more discreet, such as sebum and ear wax; rather unattractive outputs like mucus which may at times incorporate dead white blood cells and become pus; and, finally, others that seep out so slowly that they seem like growth rather than secretions, such as hair. They all have their functions, their regulatory mechanisms, their occasions. The point is that we, the head-owners, are not consulted.

Minor Secretions

The head specializes in a secretion that is produced elsewhere in the body but in far lesser quantities than in the face and scalp: sebum. It is a mixture of fats and dead cells from the lining of the hair follicles, produced by the sebaceous glands. The fats are very useful. They help to keep the skin supple and prevent it either losing or absorbing excessive amounts of water. So far, so good. Unfortunately, under certain circumstances, an excessive amount of sebum may be produced, the pores get clogged and blackheads and whiteheads form. These may get infected, as the sebum is particularly attractive to skin bacteria. The result is full blown acne vulgaris, with mildly disfiguring consequences familiar to everyone.

The particular cruelty of acne vulgaris is that it breaks out in adolescence, when one feels most defined by one's physical appearance. This is compounded by one of the body's nastier little ironies: the hormone testosterone that makes boys achingly attracted to spotless beauties is also the most important driver to the overproduction of sebum that makes them spottily unattractive. I mention this now to underline a theme that will haunt this book: that our heads are not necessarily on our sides, or on the side of our goals and ambitions that were conceived in circumstances with which the head was not necessarily designed to cope. At any rate, there can be no more intimate betrayal by one's head: it causes one to be described, and dismissed, as a 'spotty' youth. Imperfections that broadcast that one is both hungry and unappetizing, are woven into a judgement on one's callowness and ineptitude.

A lesser cruelty, due to the proclivity for sebum of inflammatory yeast, is seborrhoea. This results in copious dandruff that stands out like nits in oily hair and forms a cape of sleet over the shoulders. Sebum may be a minor secretion but, as the oily and the spotty know to their cost, it illustrates a major principle: not only happiness, but the judgement of those who matter most to us, lies in the decisions made by our flesh.

Another minor player is ear wax: cerumen.[2] It is secreted in the outer third of the ear canal and is more interesting than might appear to a Frenchman deploying his *auriculaire* (the little finger that, across the Channel, is honoured for its contribution to audiological health). It is a mixture of secretions from the sebaceous glands and modified sweat. It tastes horribly bitter as anyone will know who has chewed his nails after rummaging in his outer ear. That is beside the point. Like so many secretions, it performs more than one function: propelled by jaw movements, it washes out dirt and dust and any other particulate matter that might have collected in the ear canal; it lubricates the lining of the skin within the canal, preventing it drying up and getting itchy; and it has antibacterial and antifungal properties. Some stuff.

If you are Asian or Native American, your genes will ensure that you will have dry (grey and flaky) cerumen; while Caucasian cerumen is more likely to be wet and a more attractive honey-brown or a dark-brown. It has proved possible to track human migratory patterns, such as those of the Inuit, by looking at cerumen type. Historians have to explore many cunning passages in order to uncover history. So much of our body leaves involuntary traces, as felons know to their cost; but the thought that the footsteps that were effaced as the snow melted might be betrayed by our ears is weird indeed.

Ear wax may disable the very sense organ it is designed to protect. In the United States of America, about 150, 000 patients a week have wax removed by experts. The effect of cerumenolysis is sometimes near miraculous. It is as if an acoustic burqa has been removed and the auditory world suddenly brightens up. This is a striking manifestation of that most obvious of facts – the mediation of our experience through our senses. Hardly surprising, then, that in Shanghai's Great World Centre in the 1920s, specialist ear-wax extractors had stalls of their own.[3]

Caucasian cerumen is the colour of earwigs and this has spawned a nexus of associations. The earwig got its name from supposedly crawling into ears; a notion that is only slightly less repulsive than the image evoked by the French word for earwig: *perce-oreille*, an earwigger that nibbles its way into the interior of the head. Earwigging – 'influencing or attempting to influence by means of whispered insinuations' – is a brilliantly vivid term and it spreads its roots over the entirety of James Joyce's great night book *Finnegans Wake*, the counterpart to his even greater day book, *Ulysses*. H. C. Earwicker, the name of the consciousness dreaming the book is, of course, an 'earwigger'.

Sweat: From Acrosyringium to Dystopia

In the opening paragraph of his Malaysian trilogy, *The Long Day Wanes*, Anthony Burgess presents us with a striking image. An overweight District Officer is sitting in his humid, stifling office, looking hopelessly at some documents. The sweat falls off the end of his nose in regular drips, smudging the ink, making the challenge of governing this outpost just a little closer to impossible.

Burgess describes this dripping as a clepsydra, or water clock, marking the decline of Empire. The wild association of ideas which possesses our waking as well as our sleeping heads links this for me with a wonderful but hyperhydrotic conductor, coaxing his orchestra through Mahler's Second Symphony and taking advantage of the pause between movements to mop his brow, his neck and the nose from which the sweat had been dealing itself in regular breves, like a metronose.

Sweating seems rather infra dig for the capital organ of the body. Although it is not the first port of call when things get sweaty – armpits tend to be in the van of the rapid reaction force – it is in there fairly early when it's all hands to the sodium pump. Brow-sweat has an emblematic status, because of its closeness to the centre of the self: the Adamic curse was not 'to live by the sweat of your armpits', though this would have been more physiologically accurate. Anyway, mopped brows, saline-stung eyes, matted hair, are familiar signs of the body's attempts to correct the mismatch between heat generated and heat dissipated. This is essential to head off a threat to that homoeostasis or constancy of the internal environment which that physiologist of genius Claude Bernard identified as 'the condition of free life'. Unless temperature, along with blood pressure, hydration, serum potassium, and dozens of other 'parameters' are controlled within tight limits, the outlook for the organism is short and grim.

Sweating is not the most admired of secretory activities. It is, however, the result of some very fancy processes. Human sweat is produced mainly by so-called eccrine glands. In all other animals, apart from the more superior primates, it is secreted by the muckier and pongier apocrine glands. Eccrine glands, which are distributed all over the body, secrete nearly pure salt water. The body would on the whole prefer to sweat only water and conserve salt – and it does this more efficiently after acclimatization – but given that the two are mixed up in the places it can draw upon, it settles for a relatively dilute or hypotonic concoction. The sweat evaporates, cooling the skin and thereby reducing the core body temperature. Trainspotters may like to know that there is a loss of 0.58 Kcals of heat for every ml of sweat evaporated and that acclimatized adults may sweat up to 2 litres per hour, a rate which, if sustained, could lead to severe macroscopic and cellular pruning and potentially to death.

The eccrine glands are elaborate. Because of the complexity of the job they do, there is a division of labour within them. They have a production (or secretory) section and a distribution section. The secretory portion is a coiled tube located deep in the deepest layer of the skin, the dermis. It is here that the amount of sweat, and its concentration of salt, is controlled by signalling systems within individual cells, responding to commands conveyed by the autonomic nervous system from the brain. There is a centre in the hypothalamus at the base of the brain where thermosensitive neurones are located. These neurones fire faster when the core temperature of the body rises. Their sensitivity is upregulated by input from skin temperature receptors, which reset the level at which the hypothalamus issues the command 'Sweat!'.

The distribution duct rises up through the dermis and enters the epidermis. There it assumes a spiral configuration and acquires a lovely name, the 'acrosyringium', before it opens on to the skin surface and pours out its cooling libation. Cooling, but sometimes at a price. Physiological survival may be bought at the cost of social death. Whereas the small contribution of the apocrine gland may be blamed for most of the cutaneous basis of body odour, eccrine secretions do contribute and are caught up in the degradation of value-neutral sweat to adversely judged 'sweatiness'. Another's body odour is not only a sign that they may wash less frequently than they should – after all, sweat takes time to acquire a pong –

but also forces upon strangers a degree of bodily intimacy that should be reserved to lovers. The stranger to whom we are not attracted offends and annoys us by introducing himself into us: he gets up our nose. In George Orwell's dystopian nightmare, *Nineteen Eighty-Four*, there is a character called Parsons. He is an unthinking imbecile, fiercely and unquestioningly loyal to the Party. He is unwearyingly enthusiastic for all the Party's activities, especially those which involve parades, banners and athletics in celebration of Big Brother and the Revolution. He is almost defined by the stench of sweat he leaves in every public and private room he passes through. His pong embodies, or enodours, the utter degradation of civil society in *Nineteen Eighty-Four*.

The cultural accretions of sweat are therefore manifold. The very word that designates this life-preserving secretion is a troubled node in the semantic net. Ladies (as opposed to women or men) do not sweat – they perspire. 'Perspiration' broadcasts its distance from organic life and poor hygiene by being of Latin rather than Saxon derivation and taking a leisured middle-class three syllables. 'Sweat' is a hurried working-class monosyllable, brief as a grunt. Concern about the word may be in advance of concern about reality. Roy Porter noted how in 1791:

> *The Gentleman's Magazine* concurred in noting how vulgarity of speech was quite out: only 'the lowest class' now said 'sweat'. 'We are everyday growing more delicate', ironized its author, 'and without doubt at the same time more virtuous; and shall, I am confident, become the most refined and polite people in the world.'[4]

Well ahead of their Austrian contemporaries, anyway. According to Michael Steen, the street where Mozart was born 'reeked from the sewer running down the middle' and 'the Salzach River stank from the filth thrown in'. These, however, 'had the merit of concealing the odour of the people who never washed, but rubbed themselves clean and occasionally doused themselves with perfume'.[5]

Thermal sweating occurs over most of the skin. Emotional sweating is more selective, being confined to the palms, soles, armpits and forehead. Its mechanism is not well understood, though we are so familiar with it that 'breaking out in a cold sweat' is a standard metonym for acute anxiety. Its purpose is also unclear. It has been suggested that cooling the body permits it to burn more energy – as one might need to do in a frightening situation – without getting over-heated. This is fine, if the appropriate response is fight or flight, but not if the correct response is to stand rooted to the spot with horror at some gaffe.

For this seemingly vulgar outpouring can have the most remote, esoteric, subtle, unbiological causes. I once gave a talk on the treatment of epilepsy at an international meeting and spotted with sudorific anguish an error on my key slide. (Nobody noticed – probably because they were too caught up in another hypothalamic function – sleeping.) Stress-based sweating is certainly of little help to liars. The surge of sweat during the telling of a lie is the basis of the galvanic skin response, used in lie detection. Because it contains sodium ions, sweat increases the conductance of the skin and this can be recorded. The polygraphic record dips down like a clunked open jaw. Finally, and most esoterically, we may sweat when we are struggling to think. It is appropriate that the head's contribution to stress-related sweating comes from the forehead, closest to that place where thought is most carefully controlled (we are told), and that the other main contribution comes from the palms, which are implicated in more primitive modes of grasping, in the prehensile ancestor of apprehension.[6]

In his teasing essay on the Romans in (American) films Roland

Barthes notes how the conspirators in Wolf Mankiewicz's movies of *Julius Caesar* always appear with sweat on their faces:

> Everyone is sweating because everyone is debating something within himself; we are supposed to be in the locus of a horribly tormented virtue, that is, in the very locus of tragedy, and it is sweat which has the function of conveying this....To sweat is to think – which evidently rests on the postulate, appropriate to a nation of businessmen, that thought is a violent, cataclysmic operation, of which sweat is only the most benign symptom.[7]

We have moved a long way from the narrow loop between the skin, the hypothalamus and the eccrine glands. With stress-related sweating we encircle more and more of the world.

And, being the extraordinary creatures we are, we engage even with thermal sweating in a complete, mediated way. Recently, we have all been exhorted to look after our hearts. Regular exercise seems to hold one of the keys. According to the most authoritative guidelines (based on millions of data points, gathered painstakingly by vast numbers of biomedical scientists across the globe) the right sort of exercise, sufficient to hold off cardiovascular end-points, is such as to make you mildly sweaty. We cannot stroll away from death: brisk walking or jogging is required. 'Daily mild sweaters' live – and sweat – longer. We may use our exocrine function to calibrate the correct, or at least the minimal, dose. It is difficult to think of a more striking example of the head surmounting its own secretions to use them for the purpose of postponing that head's demise.

Saliva

Antoine Roquentin, the anti-hero of Jean-Paul Sartre's *Nausea*, is in the grip of a crisis. He has become acutely aware of the contingency of the world. The existence of the things in it seems insufficiently justified: they simply, heavily, inertly, are. Words that would redeem their meaningless thereness by gathering them up into a reassuring, weightless nexus of significations lose their grip on objects. Roquentin's own body seems like a mere given he is curiously required to be. He is a spectator of the mucky meat to which his name is attached. He examines his hand, which, palm upwards on the table, looks like a helpless upturned animal. He pointlessly, underivably, exists:

> I exist. It's sweet, so sweet, so slow. And light: you'd swear that it
> floats in the air all by itself. It moves. Little brushing movements
> everywhere which melt and disappear. Gently, gently. There
> is some frothy water in my mouth. I swallow it, it slides down
> my throat, it caresses me – and now it is starting up again in my
> mouth – unassuming – touching my tongue. And this pool is
> me too. And the tongue. And the throat is me.[8]

'This pool' is certainly very impressive. During my lifetime, it will be of the order of 30,000 litres. It has a multitude of functions, none of which will solve Roquentin's existential crisis. It lubricates the food, making it easier to swallow; assists speaking (hence the glass of artificial saliva traditionally made available to dry-mouthed public speakers); starts the digestion of starch with the enzyme ptyalin; and protects against infection by means of the various enzymes and immune proteins it contains. It is produced from the salivary glands by an extraordinary series of processes. Sweat, by comparison, is a simpleton and certainly under-employed.

When saliva enters the secretory ducts, its ionic composition (sodium, potassium etc.) is very similar to that of the blood plasma whence it is derived. During its passage down the secretory ducts, sodium and chloride are abstracted from it and potassium and bicarbonate ions are added. This provides an appropriate environment for the work of the enzymes. Some mucus is also added to the brew. The secretion of saliva is exquisitely controlled by the autonomic nervous system: the parasympathetic division stimulates the secretory cells both directly and indirectly by increasing the blood supply to them. These involve multi-step signalling processes mediated by exotically named enzymes such as kallikrein.

Impressive, but – to continue Roquentin's inquiry – is it me? We can command it to some extent. We can make it flow by thinking of vinegary chips. A commonplace until we try to think what this 'thinking of vinegary chips' consists of. An image, of course, rather than a sentence, but an image contaminated with all sorts of other things: paper, the chip-shop owner, a lit place suffused with the sense of the road where the local chip shop is located. Here we have a case of the thought of food being requisitioned to provide food for second-order thoughts about thoughts. Even so, this rather primitive stuff, unfazed by sophistication, flows fluently. Our animal mouth seems able to speak the language of meta-thought. Less reliably, we can dry up the flow by making ourself anxious, calling to mind one of our worries.

Nothing simple, then, about the relationship between the world we live in, our mind (which can requisition bits of our world to demonstrate the higher-order regulation of salivary flow) and the salivary glands. Which is why, so long as we are awake and well, we feel responsible for containing our saliva. We could almost define the arc of life as beginning and ending with drooling; or our pre-senile maturity as those years when we successfully police the liquid contents of our mouth.

To prevent ourselves from drooling, we keep swallowing our saliva. Like so many actions, swallowing begins voluntarily and continues involuntarily. We gather up the saliva with the tongue, sweep it to the back of the mouth, and raise the soft palate to close the nasopharynx so that the stuff does not pour down the nose.

The saliva hits the tactile receptors in the pharynx and this starts a series of events over which we have no control. The larynx moves up (as you can see by the upward bobbing of the Adam's apple) to meet the epiglottis which closes off the airway and prevents the saliva from entering the lungs. We momentarily stop breathing. The muscles of the pharynx then contract, propelling the splash of saliva into the oesophagus or gullet. At this point we not only lose control of what is going on but also awareness of it. A wave of muscular activity passes down the oesophagus, milking the saliva stomachwards, until it reaches the sphincter at the bottom, which opens and then closes. These later stages are orchestrated by a collection of neurones in the medulla and pons of the brainstem.

All very impressive and all very impersonal. This tiny pool is about as intimate as it gets, but it is somehow a mere lodger in the mouth. It is *our* sputum but has as little to do with our biography as the faeces in the colon. Of course, without it that biography would be different: it would include the story of a chronically sore, chronically infected mouth. And yet, unless we are in a metaphysically heightened state like Antoine Roquentin, it is perfectly comfortably assimilated into the most personal modes of being-here. This ambivalent status of our mouth-sourced and mouth-warmed saliva can be teased out by a rather disgusting thought experiment.

In his thought-provoking essay on the relationship between what is personal and what is impersonal in our bodies, and between our agency and our bodily mechanisms, 'Vodka and Saliva', Paul Broks recounts a scenario suggested by the psychiatrist Anthony Storr.

He asked us to consider how often we swallow our own saliva. We do it all the time, of course, without thinking. Then he invited us to imagine that, instead of swallowing, we spat into a tumbler. How would we now feel about sipping from a tumbler full of our own spit? It's the same stuff, but *no thanks*! Not even with ice, lemon, and a large dash of vodka.[9]

Whatever the status of our saliva in our mouth, the spit in the tumbler has, as Broks puts it, 'renounced its citizenship'.

And yet not entirely. It would be distinctly more disgusting to drink a glass of someone else's saliva than a glass of one's own, even if one had cast-iron guarantees that the saliva was pathogen-free. Even so, once it has left the mouth it has ceased to be saliva and becomes spit, irrespective of whether it is mine or someone else's. This invites us to consider the modulation of the passivity of salivating into the activity of spitting.

Acquiring the skill of spitting is an interesting example of our relationship to our own bodies and our ability to weave their spontaneous organic activities into our chosen, highly acculturated lives. The transformation of oral seepage into a missile ejected at will requires a special kind of attention to one's mouth. One has to wait for enough material to collect, shepherd it to the front of the mouth, slightly barrel the mouth, suck in the cheeks and then blow out, patting the bolus as it leaves. No wonder it is such a preoccupation of schoolboys (even more so in the era before the wide availability of electronic entertainments) and the distance one could spit a measure of one's wider prowess. (Even now, 'spitting distance' is a marker of proximity.) It is, of course, mildly rude and hence subversive. Football, a game dominated by schoolboys, is also associated with spitting. It is a common punctuation mark between set-piece movements; part of the breath-restoring pause.

The history and sociology of spitting is an immense topic and quite right too, because it is profoundly connected with so many aspects of our relationship not only to ourselves as persons and as bodies but to others as personalized bodies. The evolution of good manners can be traced, as Norbert Elias[10] has pointed out, through the etiquette governing spitting. In the sixteenth century, it was suggested that it might be a good idea to turn away when spitting 'lest your saliva fall on someone'; in the eighteenth the advice was not to spit at such a great distance that you might not be able to find where it landed in order to put your foot on it. In the late eighteenth century, it was ordained that gentlemen did not spit on the walls or furniture. Only in the nineteenth was it acknowledged that spitting at all times was disgusting. Even so, despite the advance of the spittoon, especially in America, sputum was liberally plastered about public spaces. The increasing popularity of tobacco-chewing in the nineteenth century contributed to making America 'one long expectoration' (as Oscar Wilde described it) and Dickens referred to Washington as 'the head-quarters of tobacco-tinctured saliva'.

This instinctive revulsion towards the act of painting public spaces with your personal effluent was reinforced by the recognition that spittle contained micro-organisms and could become the means of transmitting infection. When I was a medical student, I remember being told about knickerless little girls in the slums squatting on the pavement to play and acquiring tuberculous salpingitis (inflammation of the fallopian tubes) from infected sputum. But the repulsiveness of sputum goes deeper than the reasonable fear that it might be a source of infection. It is a deep revulsion for the bodies of others that only sexual desire can conquer, a matter which will be better addressed when we consider the strange bicephalic activity of kissing. For the present, let us examine a further metamorphosis of spitting.

Spitting on someone is the ultimate in insult, compared with which name-calling, however inflammatory and degrading, seems merely metaphorical. Spittle originates from the cavity where words are factored, so that spitting partakes of linguistic signs. The sign is arbitrary, in the sense that linguists use, of not looking like its meaning. It is, however, without grammar. It has the brutal immediacy of a fist on the face. It is halfway between a curse and blow. But it has a particularly sinister aspect. It is a forced intimacy, a little rape: the spat-upon is directly exposed to material drawn from the intimate recesses of another's body. The spittle is more powerfully disgusting if it approximates to the condition of sputum by an admixture of mucus and, worse, hawked up from the private passages of the other's head.

I shall resist at this stage a digression on the theory and practice of hawking, simply marvelling in passing at the bodily self-knowledge required to clear one's ENT tract in order to gob to a great height. Instead, I shall recall with shame the case of one D, a third-former, and G. D was the bespectacled, gawky, swotty, adored only child of his parents. Alphabetical order ordained that in class he should sit in front of G. G did not satisfy himself with endlessly ruffling D's carefully Brylcreemed hair with a ruler. From time to time, he would rummage in the greenest recesses of his airways – mobilizing an extraordinary skill which we all possess, of handless prehension by snorting, using a combination of nasal inward drafts and muscular movements – and manufacture a gobbet of the most repulsive kind. By dint of application not evident in his studies, he was able to send it to the ceiling. There it would stick for a while before it extended like elastic and fell on the head of the innocent D., occupied with an interlinear translation of *The Aeneid*. D. had a nervous breakdown which shamed those of us who had done nothing to help him. G. later ended up in jail for tax fraud.

The asymmetry of the one who spits and the one who is spat upon is profound: it plumbs the depth of the power relations between humans to its existential bedrock. Which makes this line in Handel's *Messiah*

He hid not His face from shame and spitting[11]

arresting as well as poignant. The image of the Son of God, with sputum trickling down his sweat-glistening, blood-stained cheek is a shocking confrontation with the mad notion of the Author of the Universe taking on the human condition. To be a human is to spit or to be spat upon. In the powerful sado-masochistic myth of Christ's Passion, the mysterious process whereby human beings transform natural events (in this case the upwelling of saliva in their own mouths) into symbols which can be actively used reaches an extraordinary climax.

This is not, however, the terminus of degradation. George Steiner[12] recounts a terrible story of a rabbi in Nazi-occupied Poland:

A rabbi in Lodz was forced to spit on the Torah scroll that was in the Holy Ark. In fear of his life, he complied and desecrated that which was holy to him and to his people. After a short while he had no more saliva. His mouth was dry. To the Nazi's question, why did he stop spitting, the rabbi replied that his mouth was dry. Then the son of the 'superior race' began to spit into the rabbi's mouth, and the rabbi continued to spit on the Torah.

This ghastly variant of the Eucharist, in which a Jew has to swallow the body of an evil-intentioned Gentile, says something about the twisted relations we have to the world in which we act out the condition of embodiment. This is spitting at its most virulently direct. By the principle that humans can make infinite use of finite resources, spitting can be reassumed into metaphor. The insult 'I spit on your grave', or 'I spit on your mother's/ancestor's grave' is highly abstract. The sputum in question is the referent of the word that refers to it: it is the pure idea of sputum distilled to its capacity to insult; and its target is the notion of a place – the grave – which stands metonymically for that which is, and should be held, sacred of a person.

To return for a moment to the spitting that was part of the Passion of Christ: it was supposedly foretold in the book of the prophet Isaiah[13] seven hundred or more years before Christ is supposed to have been born. What more extraordinary example – if it were true – could there be of the incorporation of an oral secretion into collective consciousness than that of spitting prepared 700 years – greater than the interval between Crécy and the Somme – before the head went back and hurled it at its target? This little pool welling up in our mouths, which accompanies us all our days from bubbling infancy to drooling anecdotage, can be transfigured into a symbol which enables it to serve so much human wickedness. We utilize what happens in our body for purposes no body could foresee. What other organism uses saliva to lick stamps and seal envelopes bearing such news, good or bad?

Reflections on Anglo-Mucus or Decoding the Doze

The nose, as no one who possesses one needs telling, warms and humidifies the air on its way to the lungs. While air supports life – millions have lived without love but none without oxygen – it may also import death. The nose is also therefore a necessary filter, bouncing unwanted particles and providing first-line defence against potentially pathogenic organisms by bringing inspired air into contact with mucous-coated membranes that contain immunoglobulin A. Immunoglobulins are proteins that intercept micro-organisms and prevent them from multiplying in the body. The membranes lining the nose produce mucus which contains muramidase, a bacteria-killing enzyme, also present in saliva (see earlier) and tears (see below). The mucus is then swept by very fine hair-like structures ('cilia') beating in unison, creating a

Mexican wave, passing from the nasal cavity and the paranasal sinuses towards the nasopharynx. There mucus can normally be swallowed.

By this means, invading organisms are directed away from the lungs, where they might thrive, to the acid hell of the stomach, where they most certainly will not. As you read this, you may be aware of a globule of mucus hanging in that just-about-someone's land on the border between your pharynx which you know so well and inner oblivion which you can only speculate about.

The head produces about a quart of mucus in twenty-four hours and this can double when the membranes are inflamed, as in response to invasion by rhinoviruses and coronaviruses, the organisms responsible for the common cold. We wake in the morning and know that it is going to be a bad head day. Mucus is blocking the nose and pouring out of it; a splinter of glass has lodged itself in the pharynx; the skully part of the head aches; and, though the world it sees, hears, tastes and smells is blurred, certain parts – burning eyes, nostrils chaffed by endless emunction – are highlighted. We tend our afflicted head, drawing deep on science (decongestants that dampen down the vasomotor drive to the mucous membranes), folklore (Echinacea) and advertising that mixes the two. The thinker whose head it is mobilizes knowledge to correct that head's transformed state.

The most striking transformation, which does not affect the voice in his own head, is in the thinker's speech. The comic possibilities of the new dialect – Anglo-Mucus – are endless. In literature, it is unpoetic souls who not only suffer this most unpoetic of events, but who seem somehow defined by it. This injustice reaches its apogee in Thomas Mann's great novella *Tonio Kroger* where the eponymous hero, on the deck of a ship, is exposed to the poetic thoughts of a stranger moved by the spectacle of the stars on a night crossing of the Baltic Sea. The stranger, Kroger is sure, 'has no literature in his belly', although he probably writes verses – 'business man's verses, full of deep feeling and single-minded-ness'.[14] The stranger has a cold in his nose, as is signified by his reference to his 'belancholy' and the 'evebing' and 'stradgers'. His cold mocks his poetic aspirations.

'Bucus' interferes with the passage of air. A cold therefore has most impact on nasal consonants. These are formed when the soft palate lifts, the tongue blocks the mouth and articulate air is forced down the nasal passages. Hence the rueful, ironic, acoustically self-referential observation that 'I have a code in by doze'. The layers of awareness built into that commonplace are extraordinary: the words I use to refer to my cold also illustrate it mimetically by incorporating the sounds of the consonants that are most affected by the affliction of the structure to which we are referring.

Mucus is not usually a laughing matter. We find it rather repulsive and have to steel ourselves to wipe the nose of a snotty-nosed child. (Being snotty metonymizes the condition of inadequacy and self-neglect of childhood, particularly working-class childhood.) We are offended by the spectacle of someone else picking their nose or examining the contents of their handkerchief after a good blow. 'Too much information', we feel. Snot (old Friesian, by the way, for nasal mucus) is the quintessence of the yukish. And yet, here as elsewhere, medicine conquers human repulsion and adopts a naturalistic attitude to the phenomena. Sputum pots are examined for signs of illness, evidence of diagnosis, proof of cure. 'It may be snot to you, mate, but it is my bread and butter', as one chest physician is reputed to have said.

The viscosity of sputum, particularly when it forms mucous plugs in the airways, may be a matter of life and death. A colleague of mine did his MD thesis on ways of reducing the viscosity of mucous secretions using so-called mucolytic agents. The viscosity of such secretions depends on the concentration of mucoprotein and the presence of chemical bonds between them. One agent splits these bonds and reduces their viscosity. That at least was the theory. Does it work out in practice?[15] This is what my colleague spent the best part of three years trying to determine. One approach to the problem was somewhat direct. He took samples of sputum, placed them between glass rods and measured the force required to pull them apart, before and after application of the agent. He found that the drug made the sputum 10 per cent more tractable and hence, presumably, proportionately easier to expectorate. This justified the advert for Mucodyne. It showed an elderly gentleman, with cheeks blown out like a horn-player, coughing into a handkerchief with the caption 'Great Expectorations for a Dickens of a Cough'.

The attitude of medicine to sputum and its congeners is the least snotty imaginable – an observation that gives me a cue for a relatively aseptic linguistic excursion. The word 'snotty' seems to have gone upmarket. We may follow its career with the help of the *Oxford English Dictionary*. It began meaning 'foul with snot or nasal mucus'; proceeded to 'dirty, mean, paltry, contemptible' ; and then, by a series of leaping knight's moves, became 'angry, curt, short-tempered', 'pert, saucy, impudent', and 'proud, conceited'. The snotty, from being slightly repulsive but pitiable, suddenly became a force to be reckoned with. They are supercilious, looking down their well-wiped noses at their inferiors, even intimidating, as *Webster's* quote from Dorothy Parker indicates: 'You were so snotty when I called you up, I was afraid to talk to you'.

The link between snot and poverty, however, remains. In his devastating account of life in contemporary Zimbabwe, Peter Godwin reports an episode in which a small boy comes up to the closed window of his car, begging. When Godwin doesn't open the window, 'he wipes his hand across his runny nose and writes on the glass in yellow snot: *help me*'.[16] The use of snot as writing material, the finger as a pen, and the car window as paper, is a measure of the total destruction of the nation by Robert Mugabe and his thugs. They have reduced the people of Zimbabwe to their own destitute bodies. Even so, they have not taken away their fundamental human capacity to transcend their bodies, as illustrated by this starving boy using his body to replace the implements he cannot afford.

Tears, Idle and Otherwise[17]

After earwax, sebum, sweat, saliva and mucus, tears may come to some as a welcome relief. A secretion at last with a bit of class. As clear as sweat, only odourless, tears emerge from an impeccable source: they present little challenge to the sensibilities. They are, however, deeply mysterious, and central to humanity's understanding of its self. The human world, the human condition, is often described as 'a vale of tears', never as 'a vale of mucus' or 'a sea of saliva' or 'a midden of earwax'.

Tears illustrate something that goes to the very heart of what it is to be a human being, and makes it less shocking to think that the very home of a free will is awash with merely biological events that are not chosen by anyone. When we put our heads together, we are able to fashion 'me-doing' out of 'it-happening'. Our collective heads liberate us from the tyranny of our individual heads, so that biological mechanisms are suborned to our human – sometimes all-too-human – purposes. But let us begin with biology.

Notwithstanding Lord Tennyson's poignant verse – 'Tears, idle tears, I know not what they mean' – we know exactly what most tears mean and they are not at all idle. By washing and lubricating that delicate, transparent window through which we receive so much information about the world, tears prevent the eye from being rubbed blind. Along with salt and water, they contain lubricants such as mucin and fats and antibacterials such as lysozymes and immunoglobulin. The eyes therefore are kept moist, grit-free and uninfected. Each blink draws a small amount of fluid from the lachrymal glands which are located near the eyelids: each thimbleful of night brings its shower of rain. When the eye is injured or a foreign body takes up residence, the shower becomes a downpour and we see-ish a world-ish while lachrymation tries to repair the damage or repel the invader.

This, then, is the theory of 'tearing'. So far, so straightforward. And then it gets very strange – and distinctively and exclusively human. While other mammals may lachrymate in response to extreme pain, as well as to launder the eye, only humans cry or weep as an emotional reaction. In such cases, the lachrymal glands contract to increase their output, in an ocular ejaculation; and while some of the output will take the usual exit into the nose, stirring up the mucus and making it run, the ducts may be overwhelmed and tears trickle down the cheeks. In addition, there may be sobbing: contortion of the face, with convulsive expiration and non-linguistic vocalization.

There are all sorts of explanations as to why man is, uniquely, the animal that weeps.[18] The most plausible is that humans are singularly immature. A form of crying is seen in other animals: 'a separation cry', a call for help to an absent mother. In humans it assumes much greater importance because of our undeveloped state at birth. 'The sense we are born with' is grossly inadequate to make us safe and continues to be so for an extraordinary length of time. The meadow-wisdom of a two-year-old sheep is beyond any street wisdom imaginable for a two-year old human. The reasons for this go deep: so much more of our lives is governed by conscious decision-making guided by abstract knowledge than is the case with sheep. The latter are directed largely by instincts and mechanisms. We live in a world fashioned in several million years of cultural evolution that our untutored bodies have not caught up with. The separation cry is consequently not only more important, but also important for longer, in us. We cry to communicate our distress, our lostness, our well-founded fear, to our mother.

So far, *fairly* straightforward. The howling component calls through mother's inattention or the forest's or bedroom's darkness. Children who cry are more likely to survive; evolution will have selected for cry-babies – whose vocalizations harpoon their mother's hearts – over stoics whose silence may well be fatal. What is more, in line with the usual principle in biology of two (or more) for the price of one, crying stimulates lactation in the mother: one howl requisitions safety, warmth, food and lodging.

But this does not explain crying in adults who are no longer potential 'babes in the wood' of a hostile world, especially since adult crying most often takes the form of silent weeping: tearing without vocalization. The biologists offer an explanation for this. Adult humans are cry-babies because they have 'neotenic' brains, in which infant anatomical and physiological characteristics persist into and beyond maturity.[19] The anthropologist Ashley Montagu has claimed that crying – and laughing, which we shall visit in due course – are the most prominent childish traits that we preserve throughout our lives.

This, then, is the biological *donnée*. The tears are a sign of distress – usefully located in the part of the body that is most easily seen (a seepage of sweat between the toes would be less useful) – and they draw attention to themselves with the aid of sobbing. But, as is invariably the case with humans, the biological given is just the start of a long journey. Humans are, collectively and individually, like William Webb Ellis, the inventor of the game of rugby. In the middle of a football match, he took up the ball and ran with it. And this action then set in train a cascade of events as humans followed a uniquely human trajectory. Further down the track we find all sorts of strange phenomena: committees meeting to regularize the rules, snobbishness about professional versus amateur status, arguments about the respective merits of Sir Clive Woodward and A. N. Other for the post of chief coach of the national rugby squad. The biological, therefore, takes us only so far. It explains why, for example, the audience is moved to tears by a perfect performance of Schubert's String Quintet in C Major rather than (though it addresses heart and soul through sound) an upwelling of ear wax, so they are left dabbing at trickles of honeybrown or grey-brown cerumen coursing down the sides of their heads. But it leaves a lot of unsolved mysteries.

Here's a semi-biological one. Emotional tears (for example, those provoked by Tennyson's line 'Dear as remembered kisses after death') are richer in manganese and protein than those prompted by pain elsewhere in the body or by chemical irritants or than in the regular windscreen wipers. It has been suggested that emotional tears may get rid of excess stress-related toxins but there is no evidence to suggest that someone who cries tears in response to emotional stress will develop an internal chemical imbalance requiring correction.[20] Besides, eye-pissing seems a rather clumsy way of dealing with the problem: taking a scalpel to chop down a tree; or trying to pay off the national debt by collecting milk-bottle tops. The kidneys are better equipped for the job. Emotional crying seems designed to deal with toxins of the soul rather than those of the body.

Even here, though, biological thinking seems inescapable. One of the mysteries of crying is that we may cry with joy as well as with grief or pain. It has been suggested that 'joy or relief liberates us to recognize and react to stored up pain and sadness, so that our tears are actually a reaction to sadness, not happiness'. Crying could therefore 'be considered as a kind of psychic homeostatic mechanism, returning the body to an emotional equilibrium which has been upset'.[21] Well, yes and no. 'Homeostasis', as we noted when we discussed sweating, is the phenomenon, crucial for the viability of the organism, whereby the internal environment is kept constant. This doesn't seem to fit emotional crying: the equilibrium in question is only tenuously related to, say, the maintenance of physiological parameters.

Time, therefore, to leave biology, in which crying may have its roots, and look towards human culture, in which it unfolds its leaves. The way out of biology is hinted at in the passage just quoted, about tears being the result of joy liberating us to recognize and react to stored up pain and sadness. Tears that have been withheld during an ordeal are released when we are relieved or rescued. Children, Schopenhauer points out, cry when they are comforted: it is the *idea* of the suffering rather than the suffering itself that makes us weep. For the emotions that prompt tears may be highly sophisticated. Our emotions – which occupy much of our waking lives – are propositional attitudes; imaginings that are interwoven with words spoken to ourselves, spoken to others, or in imagination spoken to others. The crying animal talks himself or herself into tears.

The tears may be judged by the quality of the emotional thoughts that lie behind them: self-pity, pity for others, sympathy, compassion. At the bottom of the scale, we have the tears of a child angry that another has been preferred over himself; or that an ice cream has been denied him; or that it has dropped on to the sand and is not to be replaced. This is where crying may broadcast the bad weather of an undeveloped soul and it takes the form of whingeing, grizzling, blubbing, boo-hooing, snivelling, and so on. There are tears of vexation and frustration that bespeak our impotence. They may be accompanied by physical actions such as the gnashing of teeth that bite themselves for the want of an accessible or appropriate external target. There are tears of grief that we all respect because they draw upon the sources of suffering to which we are all vulnerable, however choreographed they might seem to be. These are tears we want to share. Indeed, we worry for the bereaved one who does not weep: we fear the effect of all those undischarged emotions, those unreleased toxins. The higher modes of crying tend to take the form of silent weeping: they do not demand help or even attention from others. They prefer solitude: they do not call out from the darkness of the forest to the lost mother.

Crying empowers the powerless. The mixture of accusations and tears, both showing and telling hurt, is overwhelmingly effective. Many have stayed with partners they have tired of because they cannot bear the judgement passed on them by the spectacle of naked suffering. Saline may not be an argument, sobbing not a justification, but they make arguments seem feeble, dry, calculating things. Crying may enable the weeper to regain the initiative: the argument cannot continue until the sobbing stops, and the sobbing can start again at any time. The one who cries moves the discussion beyond words, calls time out, and leaves the other helplessly waiting, or raising the voice to rise above the sobbing, or raising a fist.

Tears may be requisitioned for ever more abstract purposes and directed to objects remote from the primary functions of saline, mucin and lysozymes, and the separation terror of the lost child. The actor Emlyn Williams wept when he learned that the last Cornish-speaking monoglot had died. Simone Weil cried for the hungry of China. Tears may be mediated by the idea of one's self as sensitive. Thanks to Jean-Jacques Rousseau, and other persuasive leaders of fashion, tears became valued as signs of rather special emotions. The Man of Feeling – who could give way to tears without rotting his moral fibre or impairing his ability to provide leadership, or deliver protection and economic support to the weaker sex – was elevated to the status of a higher being. Novels were written in order to jerk tears by means of idealized sorrows: goodness confronted by wickedness and injustice; true love seeking expression in the face of an unfeeling world of convention and calculation.[22] They were special tears, of course: the ones rich in manganese and protein.

Ocular incontinence is not always valued. Withholding or rationing tears may be seen as a mark of deep sensibility. Spartan values are widely distributed through societies, especially male-dominated ones. Some cultures, such as the Minangkabau people of Indonesia, forbid crying altogether. And we expect the statesman to wipe away a single tear with a leather-gloved index finger as he places the wreath at the cenotaph rather than work his way through box after box of Kleenex. The pathos of *Boule de Suif*, the roly-poly lovely prostitute in Guy de Maupassant's famous story, who is used by the aldermen on their day out and then ignored by them on their post-coital journey home, seems all the greater for the singularity of the fat tear that rolls down her cheek at the end of the day. And weepers may provoke suspicion. There is the feeling that weeping is a style of communication that can be used at will. Critics of lachrymation reach for the distinction between true sentiment and a sentimentality that enjoys itself. The latter prefers the emotional orgasm to the boring business of doing anything about the wickedness that prompts the tears. Examples from literature include the countess in Tolstoy's story who wept copiously at the suffering on the stage while showing indifference to her coachman, freezing to death outside, waiting to take her home. And, as auto-critics, we feel ashamed of ourselves, when the lights go up, for weeping at a weepy film; for flash floods of shallow emotion.

We live so much in ideas, that our tears are more easily moved by fictions than by realities. Hamlet is appalled that the Player King can weep so readily over Hecuba, while he, whose father has been murdered, and whose mother has contracted an incestuous marriage with his uncle, the murderer, remains dry-eyed. Rousseau himself fulminated against theatrical tears:

> At the theatre we cry at the tragedy lived out on the stage, and our emotional response creates a warm glow of self-satisfaction. Then when we leave, we dry our eyes and carry on as normal; perhaps we behave even worse than before. The stage, thought Rousseau, turns us from agents to witnesses, and the desire to fight inequality and injustice drains out, too, with our tears.[23]

Tolstoy and Brecht could not have put it better.

And yet the spectacle as a source of delicious sorrow may be a constant in humanity. Man, as the philosopher Alfred North Whitehead said somewhere, is the only creature who cultivates the emotions for their own sake. The weeping eye enjoys what it sees, the *theoria* that is the object of contemplation. Tragedy, Aristotle tells us, through its perfect structure and proportion, gives us pleasure even as it invokes fear and pity. We experience an ideal pity, an ideal terror and our souls are purged, temporarily, of those emotions. Or by seeking them out for their own sake, we seem to conquer them; to possess a piece of the world that most profoundly possesses us. We see the tearworthy with dry eyes, and so we experience clearly those events which in real life we suffer with clouded consciousness served up by a blurred gaze. We are spared, too, those consequences of weeping that remind us of its lowly, non-metaphorical origin: the snot flushed out by tears, the smudged make-up, the swollen eyelids.

There are other ways of dealing with tears. The modern man of designer sorrows, acquainted with chic-grief, wears shades. These, as Roland Barthes pointed out, conceal our eyes from the gaze of others but, by doing this so blatantly, broadcast what is concealed.

The darkness of the glasses casts light on the emotional state of the unhappy lover. The black discs stand for the darkness of the soul, the sleepless nights it has passed. '*Larvatus prodeo*: I advance, pointing to my mask'.

Perhaps the last word should go to Racine's Phaedrus:

> *Je ne pleur pas, madame, mais je meurs.*
> (I weep not, madam, but I die.)

As the bright-eyed marmoset, far from warmth, and the babes in the wood, and the little hominid in the forest, all know: we must cry out to one another or die.

First Explicitly Philosophical Digression:
Being My Head

In our preliminary engagements, we noted how little of our head was available to us at a given moment of our lives; and how quite a bit of it is never available to us at any moment of our lives. If this seems worth thinking about, and worrying over, it is because we assume that, if we are in any sense to think of ourselves as being identical with our heads, it is at the very least because we colonize it with immediate awareness. The less of it we colonize in this way, the less of our head seems to be ourselves.

Is this perhaps a misconception? Certainly, it makes the head-that-is-me seem rather small and distinctly volatile. It entirely excludes the brain, which is numb to itself, most of the cranium, and a good deal of the extracranial structures for most of the time. What's more, I don't identify particularly strongly with quite a lot of the material that lights up when something happens to it. The complex of a trickle of saliva and the back of my teeth for example – or more precisely my saliva-bathed teeth – doesn't seem terribly like a particular 'I' when that 'I' focuses on it. In fact, the harder the 'I' looks, the less there is to find that seems to be the 'I', to be what the 'I' is.

This may be a clue. Attention itself externalizes what it attends to. What I attend to is not me, the attender. The harder I attend to, say, my mouth, the more firmly it is placed at the end of my attention as its object. Experienced tissue, made the focus of attention, has the character of something I *stand in relation to*, rather than something that I *am*. Of course, if I were to search other parts of my body for flesh that I am, I would come back even more empty-handed. Feet, spleen, the colon are even less promising candidates for what I am than those stretches of my head – mainly in my mouth – that are flushed with my immediate self-awareness.

This kind of thinking provokes desperate responses. The commonest mode, and the one that has the longest and most interesting history, is to distance one's self from one's body altogether: to conclude that I am not my body, that the true me is a non-bodily something that happens to lodge in my body. That something does not share the fate of the body, or may do so only temporarily. It may be a soul or merely a mind, a thinking substance. The soul is a hybrid born out of a collision between morality, theology and metaphysics. The mind is a seemingly less murky notion than the soul but definitely not simple. It seems to encompass experiences of the body, experiences of the world, emotions, intentions, and (something woven in with them all) thoughts.

Distancing the self as soul or as mind from the body raises all sorts of unanswerable questions about how I came to be caught up with this body; how I interact with it; and how I experience it and the world through it. What is more, the soul or mind seem rather impersonal. It is difficult to see how, out of their very general properties, anything as particular as a particular person could arise.

Even Descartes, the most famous proponent of the identification of me with thought (a thinking substance) was not entirely consistent:

Nature also teaches me by these sensations of pain, hunger, thirst etc that I am not only lodged in my body as a pilot in a vessel, but that I am very closely united to it, and so to speak so intermingled with it that I seem to compose with it one whole.[1]

However, he still felt that 'the real me', the essence of René Descartes, was *thought*. In his Sixth Meditation, he asserted that 'all these sensations of hunger, thirst, pain, etc are in truth none other than certain confused modes of thought which are produced by the union and apparent intermingling of mind and body'.[2] Unconfused, however, they would be 'proper' thoughts – impersonal things like sentences. No wonder the German philosopher Georg Christoph Lichtenberg suggested that Descartes' mind-self was impersonal: more like 'It thinks' or 'There is thinking'. It has even been suggested that Descartes' famous argument could be glossed as 'It thinks – therefore I am not'![3]

Another response is to suggest that the failure to identify myself with some piece of my body, or even my head, is not accidental. The elusiveness of the 'I' is, according to the Oxford philosopher Gilbert Ryle, 'systematic' because it is in the very nature of the 'I' not be identified with a particular part of the body – or indeed, a particular anything – not even a series of experiences, a collection of traits, a succession of events, never mind a piece of matter.[4] Sartre was even more radical, identifying the 'I', what I am, with nothingness. He came to the paradoxical conclusion that the 'I' is what it is not and is not what it is.

There is a theory behind this: a theory of consciousness, or the for-itself, which is central to the Sartrean self. The for-itself is nothingness, 'a worm at the heart of the plenitude of Being' which enables Being to 'exist', to stand in relation to itself – to be about or of itself.[5] Nothingness is slippery stuff and not altogether convincing as a player in the world, in human affairs. The '-ness', which raises nothing to a second order, seems to curdle it a little, giving it an almost substantial reality, seems like a grammatical trick.

And so we need to look elsewhere for some way of dealing with the rather extraordinary state of affairs whereby my self, my sense of what or who I am, has a very insecure grip on any part of my body, even the head that is closest to being me.

One way is to think of the body with which I am identified, and as far as I am identified with it, as a kind of *assumption* or presupposition. That I *am* this body (or certain parts of it at certain times) is presupposed in other relationships I have to my body. It is their starting point and the point where those other kinds of relationship, which we shall discuss at intervals in this book (suffering, owning etc), converge. We may imagine the newborn infant simply being its body, in the sense of living its experiences, but not having an explicit relationship to it. It gradually comes to realize that this body is itself. A blush of awareness evolves into a many-layered sense 'That I *am* this body'. At this point, a gap opens up between the infant and its own body as it realizes that this body is its own.

And yet the body still remains as a kind of profound assumption – in two senses. We assume our bodies in the way that Jesus Christ is supposed to have assumed human flesh when he came down to earth as a human being. It is a form of donning that goes deeper than any other: it is donning *one's self*. Doff it at your peril. And we assume our bodies in a second sense: as that which we can and must take for granted; as the background assumption from which we operate in living our particular lives.

Our bodies, in short, are the existential ground upon which we stand: the arm of the bowler, the legs of the walker, the head of the onlooker. Descartes' famous *cogito* argument – 'I think therefore I am' – was superfluous. This is obvious, of course, from the fact that there was nothing in its conclusion that was not already present in its initial premise. He might just as well have said 'I am therefore I think' or 'I therefore I'. Which was why the conclusion, though it proves less clear on close inspection than at first sight, was so secure. What he was *not* doing was successfully completing a journey from uncertainty about his own existence to a reassurance that he exists. We don't really have to *prove* to ourselves that we exist – how could we? Our existence is the ground on which all proofs stand. What Descartes was really doing was reporting the spread of a realization across his thinking body that he *is*; iterating the existential assumption, the assumption that he exists, an assumption that went live as soon as he assumed his body as himself.

There is still, however, a gap, the gap we just referred to; so that whatever bit of my body I attend to, becomes something I attend to, not something that I am. When the infant eventually matures (or some might think *im*matures) into an introspective philosopher, the latter attempts to cross this gap: to find an immediate identity in the relations it has with the parts it is attentively aware of. But the unquestioning oneness with the body, prior to the opening of explicit relationships with it, cannot be retrieved. Other relationships have taken over. It is these relationships that we shall examine at intervals, if only cursorily, as we try to get our head around embodiment.

A couple of further points before we leave this knottiest of themes. First, I have assumed that any search for identity with the body should begin and end with the head. I implied as much earlier. But there are times when, if not the centre of the self, at least the centre of embodiment seems to be found in other places. When I have gut ache, this is certainly the case. And I often find that the pleasure of thought or memory or anticipation is lost as the space that these would normally light up is occupied by nausea and malaise. When I am engaged in effortful activity, I diffuse a little into the trunk and limbs that are involved. And when I am grieving, or anxious, the part of my body that I am seems, again, to be somewhere around my stomach. The second point is that it is probably a mistake to think that our location within our body is something that is determined entirely in-house. That with which others identify us – and that *by* which others identify us – is a potent influence on what or where we feel that we are. In my own case, the recognition that my face receives seems to suggest that I am it and it is me.

On the other hand, the head seems transparent, even absent, when we concentrate on thinking. Then it is these thoughts that seem to be what we are.[6]

chapter three
The Head Comes To

From far, from eve and morning
And yon twelve-winded sky,
The stuff of life to knit me
Blew hither; here am I.

A. E. Housman, A Shropshire lad, xxxii

The Head's Construction

We rather take our head – its existence and all its peculiarities – for granted. This is not altogether surprising. After all, taking the head for granted – presupposing it, assuming it – is central to being what we are. But it does make it rather difficult for us to imagine that there was a time when our own heads did not exist; that they came into being in a particular year. We know this as a fact, of course – the kind of fact we would not relinquish without a pretty fierce struggle. We know that for millions of years the world was unoccupied and unobserved by our heads. The world was just as busy, as rich, as serious, and as full of its own importance.

The construction of a human head – the journey from conception to birth – is one of the most remarkable events in the biosphere. It includes the construction of the brain which is, we are repeatedly told, the most complex object in the world. (The fact that the notion of 'complexity' is a product of a collective of brains should make one a little suspicious of the objectivity of that claim!).[1] And then there is all the kit that has to be fitted around and attached to the brain: eyes, ears, nasal passages, olfactory nerves, skull, neck muscles, salivary glands, lips, and so on. All of this takes place without anyone intending or doing it – least of all the head in question. (This raises some interesting questions about the comparative value of conscious agency over mechanism and makes one wonder why human consciousness should be seen as giving us such an edge over the competition. If you want to do something difficult, let mechanism rule.)

It would be easy at this stage to drown in facts. The whole business of putting the outline of the head and neck together takes five to six weeks. At Week 3, we have a three-layered disc of cells. This multiplies and folds over itself so that by Week 8 we have a brain differentiated into cerebral hemispheres and brain stem and so on; cranial nerves connecting the brain with sense organs and muscles; highly developed sense organs (the eyes have lenses, the middle ear has a trio of bones to transmit vibrations); a recognizable face, equipped with tongue, mouth, nose; a sophisticated set of muscles that will control mastication, facial expressions and move and support the head; and a cranium and organized facial bones. Just like that. While these momentous events are taking place, developments that may determine viability or non-viability and, if viable, a lifetime of happiness or grief, of achievement or failure, the mother may not even be aware that she is pregnant.

Even if she is aware of the unfolding within her, she cannot guide it, only wait for the result. The embryonic head is self-organizing. This involves three fundamental processes. First, there are repeated cell divisions turning a handful into billions. Secondly, there is the differentiation of cells into various types: skin cells, nerves, the capsule of the lens, and so on – all wonderfully fitted to their profoundly different purposes. And finally, and most mysteriously, 'morphogenesis'. This term designates the processes which control the distribution of cells so that they are organized into tissues, organs and overall anatomy. It ensures that all the cells of my ear lobe are clustered together; that the various parts of my eyes are wired in properly; and that eyes, ears, nose, cranium, facial muscles and so on have the right shape and the right relationship to each other.

It is still very far from clear how this happens. Scientists have identified certain 'morphogens' as key to the process. Morphogens are, as their name indicates, shape-generators. They are soluble molecules that diffuse among the dividing cells and, on the basis of their concentration, tell the cells what they should differentiate into. This ensures that skin cells are manufactured where the skin is to be and not where there should be bone; and that light sensitive retinal cells end up at the back of what is to become the eye, not in the middle of the cochlea of the ear.

Brighter heads than mine are teasing out the other factors that enable genes to be expressed in different ways in different parts of my head so that the right cells are located in the right place and connected appropriately. There are so-called transcription factor proteins that interact with the DNA of different cells and determine how that DNA is expressed. These transcription factor proteins are themselves coded for by master regulatory genes that can activate or deactivate transcription. In addition, there are cell adhesion molecules that determine whether a cell attaches itself to other cells and settles down or whether it continues migrating. And that's just for starters. Millions of detailed instructions are necessary to create your head that is currently trying to get itself around these strategies.

Sometimes the mechanisms seem very simple. Take the instruction which ensures that the eyes are the appropriate distance apart. There is a protein, the product of a gene called Sonic hedgehog homolog (SHH), which tells the plates of bone (the 'facial primordia') that give rise to cheekbones, forehead and nasal passages and so on to grow or not to grow. If the facial primordia grow, the sockets of the eyes will separate. If they do not grow, the orbits will not separate so that, in an extreme case, the growing head will have a single eye like the Cyclops. If they grow too much, the eyes will be pushed around the corner towards the ear.[2] This apparent simplicity is, of course, deceptive, like the Yes–No of a switch in a billion gigabyte computer.

So far as this head is concerned, all of the above took place in the uterus of Mrs Mary Tallis in the spring of 1946. Which makes it rather remarkable that I was speaking to her only last night, reminding her that she had been my first home and joking that her hysterectomy a few years ago had been like a post-war slum clearance. We talked about the accident of my existence, and the accident of hers, and the consequent accident of this conversation about the accident of our existence. It struck me then as very odd that you use the same language to speak to someone in whose body you have grown, from whose tissues you have requisitioned the means to life, as you do to a chance-met stranger in the street.

It is strange that this head exists, because this head came about as the result of a myriad of processes that I know nothing about; that it arose in a particular place in the world and emerged to engage with the world from a particular place; and that this all happened at a particular time. What is the force of the 'particular' here? It is this: the head that I more or less am is just one object out of many millions that I know or know of; that the head is a grain of sand in the infinite heap of the things the head knows of. I am reminded of the anguished cry of one of the characters in a play by Christian Dietrich Grabbe: 'Only once in the world and of all things a plumber in Detmold';[3] that is, to have been assembled in order to know a universe and then to be just one small thing in the universe that one knows.

On the fringe of my conversation with my mother was an intellectual and existential tension that I would like to tease out. At the bottom of this tension is acknowledgement that this thinking head is the beneficiary of a multitude of accidents that have determined its exact form, its trajectory through the world, and the very fact that it exists. The fact that it exists, so that it can be amazed at its existence, lies at the end of a series of utterly unfated happenings. The first is that there is a world (that there is something rather than nothing). Secondly that carbon – the basis of life – came into being. This is highly improbable according to the highly improbable laws that shaped the evolution of hydrogen atoms. Thirdly, that carbon should have come into being on a planet where there is the right balance of light and water. Fourthly, that life should have evolved, in the way that it did, out of the dances of carbon with a widening circle of partners. Fifthly, that multi-cellular organisms should have grown out of molecular pre-life; that vertebrates should have crystallized out of multi-cellular life; and that hominids should have awoken out of nature. Sixth, that a hundred thousand generations of hominids should have survived just long enough to secure the existence of the head's parents. Seventh, that a million contingencies should have brought those partners together; and that, when they were together, they should have made love at a particular time to secure the union of the egg and sperm which formed the genomic basis of the phenotype that is this head. These are just the headlines of the odds stacked against this head thinking of the odds stacked against it. No wonder my gestation in the womb of the universe was some 15 million years; and even less wonder that it will nurture me for a rather shorter time. Looking back up the chain of events from the Big Bang to your little bonce emphasizes how you are an accident whose happening, for all but a few months, nothing and no one was awaiting.

Readers may feel that there is something a little factitious about whipping up a frenzy of amazement in this way, approaching one's self from the standpoint of general possibilities. After all, looked at from the Space of General Possibility held open by knowledge, any (exact) actuality has almost zero probability: you become the lucky instance of a general trend, which belongs to broader general trends, which belongs to a broader general trend – an so on. In other words, the thoughts you mobilize to astonish yourself with your head's existence – and to make that head's pupils dilate – are the result of locating your head in a Space of Possibility that is a construct of the head – or, rather, of all our heads put together.

Several decades ago, physicists began talking about the Anthropic Principle, to try to explain why certain basic conditions were necessary to make life possible. If the balance of forces within the atom were just a little different (by a few per cent), the universe would have been deprived of free protons. There would be no hydrogen and consequently no stable stars and no water. If the ratio of gravitational to electromagnetic forces were very slightly different, all stars would be either blue giants or white dwarfs. Our beneficent sun, and the planet earth, or anything like them would be impossible. Life would not have arisen; nor, as a consequence, consciousness and knowledge. The universe, in which the laws of physics are observed, however, will necessarily have laws and constants that make it possible – indeed inevitable – for there to be conscious, knowing human life, in particular conscious physicists. If there are other universes in which these laws do not hold sway, they can be neither observed nor known, though their bare existence can be postulated. Likewise, in order for me to be amazed at the improbability of my head coming into being, my head has to exist. My astonishment at my existence depends upon my existence.

Any statement we make about ourselves – such as that 'Raymond Tallis exists in the twenty-first century' – is contingent.

We can imagine circumstances where he does not exist and this head could not look at this head in the mirror. What we cannot allow ourselves to imagine is Raymond Tallis saying (truthfully or sincerely) 'Raymond Tallis does not exist' or 'I do not exist'. My astonishment that I do exist – when I might not have existed – is half way between these two positions. It is possible that I might not have existed but in order to entertain this possibility I have to exist.

What is this all about? It is about the fact that the head, once out of the uterus, comes to itself and sees *that* it exists, and sees, what is more, that it is a something in the world which it gains access to through its senses and something that goes beyond its senses – namely knowledge. We knowing heads are acutely aware that our heads have no necessary existence; that they are small things in a huge and careless and forgetful world, that mislays everything that comes into being. This coming to itself is a long and tortuous journey.

Coming To (1): Cognitive Growth

In his wonderfully funny review of *The Lore and Language of Schoolchildren* (1959), by Iona and Peter Opie, Philip Larkin makes a savage, and largely justified, attack on the children we have all been.[4] 'It was that verse', he wrote 'about becoming again as a little child that caused the first sharp waning of my Christian sympathies.' Larkin proceeds to describe the condition of early childhood – no 'money, keys, wallets, letters, books, long-playing records' etc and, worst still, 'having to put up… with the company of other children, their noise, their nastiness, their boasting, their back-answers, their cruelty, their silliness'. What makes the young child unbearable, is that it 'it hasn't been to work all today, and it hasn't got to work all tomorrow'.

This, however, does not mean that it has not got work to do.

Although it doesn't look like it, childhood is serious work – the condition of all other work. The growing child has to construct a world – a world that is in some sense inside its head (of which more later), so that wherever its head is, that head is located in a world that makes sense to it; so that whatever happens, it knows what and who it is, where it is, where it is coming from, where it is going to, how it got here and how it will get there. This massive task of world-building, of the head coming into its own epistemological estate, is a network of long, involved journeys. Here are a few brief comments on some early, fundamental developments.

a) *We acquire our bodies as our own*. The intuition that 'I am this body' is the foundation of the self and of the world and of the sense of being a self in a world. This blush of recognition evolves: it intensifies; it colonizes more of the body; it becomes structured into a self and the self expands into an inner kingdom.

b) *We gain control of our heads*. This is an important landmark: when the child 'sits up and takes notice', and sitting in its pram, harvests sense experiences from hither and yon.

c) *We discover the remainder of our body*, beginning with the most conspicuous outlying parts – the hands and the toes. The hands are gazed at in wonder, as they are slowly grasped as things by which the world may be grasped; and the toes are sucked with more undiluted attention than the pipe of Sherlock Holmes wrestling with a three-pipe problem. The unique way in which our heads relate to these parts of our body lies at the heart of the unique way we relate to the world beyond our bodies. That we became the only animals who journeyed to the moon was prefigured in the manner in which we explored our bodies and discovered our toes. The importance of this phase has recently been acknowledged in proposed government policy: monitoring of child development will include noting when the hands and toes are discovered; when the head learns to suck them. This is not as strange as it sounds: appropriation of one's own body as one's own is a necessary condition of citizenship.

d) *We discover that the world is made up of objects that are independent of our experiences of them.* We have the intuition, which could not be founded in sense experience, that things exist even when we cannot see, taste, hear, smell or touch them. Among these objects is our mother, though it will take us a long time to be sure that her temporary absences are not permanent farewells, that occultation is not death.

e) *We leap from the intuition of the independent material existence of other people to the intuition that they are aware of me and aware of the world, and that they have a different angle on things.* We start pointing things out to them, sharing our world with them.

f) *We have a sense that things in the world are connected:* that there are causal relations, as a result of which we can bring about Event B by bringing about Event A; that there are patterns in what is and what happens, that the world is something that can be more or less predicted, more or less understood and more or less controlled.

All of this is achieved in less than a year. Some important skills are built into the starter pack: newborn babies have an exceptional ability to discriminate speech sounds, so that they can recognize and show a preference for their mother's voice by the time they are three days' old.[5] But the scale of early achievements should not be underestimated. Computers can be programmed to beat grand masters at chess but the art of making sense of the light, and hence of making sense of the world, by a two year old is not matched by any other living creature or by the most powerful Cray super computer. More changes occur in the first two years of life than at any other time period.

Under the po-faced term 'cognitive development' is gathered the expansion of the child's awareness into a realm of things, into space and into time – the next moment, the next week, the next year, 'my future', 'the future of humanity'. Time gets tabled and the shoreless 'now' of the infant is increasingly articulated into an ever-expanding future. A modest expansion of the head is accompanied by a massive expansion of its world – of the Space of Possibility in which it locates itself.

None of this would be possible without steps a) to f) and with the exception of a) these stages are unmatched by our nearest primate kin whose skills are greatly exaggerated.[6] These lay the ontological, metaphysical, epistemological ground floor upon which our being-in-the-world is built. It is on these foundations that we acquire a system of linguistic and other signs that refers to things that may or may not be present; construct a lattice-work of factual knowledge; acquire the customs and practices appropriate to our cultures; have a feeling for the general principles of everyday ethics and everyday physics, and a sense of how the natural and artefactual world works. In short, these give us the springboard to move from the nursery-wise infant crawling on the carpet, detained by a speck of colour, to a worldly-wise adult deploying a million modes of know-that and know-how to pass from place to place, institution to institution, from country to country in pursuit of seemingly purposeful intermediate ends.

Coming To (2): New Every Morning

Capturing waking is like trying to recall one's recovery from the confusions, inadequacies and incompetencies of childhood. Awakening should be an easier story to tell. For a start, it's a bit nearer in time. We woke only a few hours ago, while it is years, or decades, since we left childhood. What's more, we repeat awakening, or simulating it, on a daily basis. Even so, capturing emergence from sleep is more difficult than writing a *Bildungsroman* because the changes to be documented are more profound. The problem in both cases, however, is similar.

First, the continuity of the change makes the process as formless, and hence as elusive, as fog thinning to mist. Secondly, observation of selves as with observation of elementary particles interferes with the observed. You can't creep up on your own undarkening with binoculars, pen or dictaphone. The act of recording takes you a long way away from the state that is to be recorded. Thirdly – and most importantly – you are no longer the person you have changed from – the one who was in the process of waking up. And finally, 'coming to' is about coming to a world as well as to oneself.

When this process is slowed down, or repeatedly interrupted and resumed (two drifts upwards and one drift down), as after a drunken night out, or a minor operation, or when you wake outside of your usual bed, 'coming to' has the form of 'piecing together' yourself and your world. This should be a little easier to catch hold of than ordinary awakening. But it isn't, because awakening is a reconstruction of a self-in-world – you cannot separate returning self from returning world. When, for example, you recognize the curtains, the curtains, being recognized, become flags in the centre of a world. They confirm where you are. This leads seamlessly to your room, to this particular day, to that worry, and that hope. Where you are says who you are.

That's too quick. Try again, more slowly. What sort of things do you first notice when you wake? Wrong question. By the time you have arrived at recallable particulars, your waking is well advanced. The crick in what turns out to be your neck, the warm body next to your own (that itself assists in turning your meaty heaviness into 'my body'), the weight of the sheets, the daylight that turns out to be full moonlight, are late products of the gradual discovery, or construction, of an outer world, from which the self is separated, and of a self separated from the inner world – that fake world of dreams, unchallenged by a rival reality dancing to its own tune (to which you, too, will have to dance), where the inner self, unfolding according to its own laws of growth and decay, had been sovereign...

It is not weakness of will (or not solely weakness of will) that prevents us from recalling and articulating ordinary awakening but the unavailability of a point of application for that will. We don't know how, or where, to attend. Apart from the already-noted fact that the would-be rememberer and he who(m) he would remember are not quite the same person – so that we are partially denied the privileged inside view we have on ourselves – this inability to recall anything precise or particular about awakening is connected with its essential nature. Awakening is global. It is not, therefore, a question of this, and of that, of an object here, and a memory there, a light yonder and a shadow nearer, but of a dawning over the whole. How, then, can we reasonably expect to grasp (never mind record) the global transformations of our self-light and its demisting to a day-lit world?

Instead, we shall have to satisfy ourselves with reminiscences of the successive realizations of the pretty-well-dawned. You recognize the curtain and hence the world to which it belongs and thereby cognize yourself. The warm leg next to yours marries you to the world just crystallized out of murmurous blurs. There is a wanting to scratch one's nose and to go for a pee mildly disagreeing with the disinclination to wake oneself completely, to shed the nimbus of drowse.

This is an instructive instance of the self not wanting to become the full self, exposed to a world which makes such demands and upon which it makes equal demands, especially as a blackbird's song says it is spring and his bright bill ignites childhood memories that ferry the drowser into a state where a discredited dream recovers its credence. Until, that is, the ferry of lapses strikes a rock: a startled realization that you are you and you are *late*. Sudden propulsion from the dissolved to the particular: upright, alert, agenda-laden; fully en route from the unbounded realm of sleep to the fumbled cufflinks and the snatched coffee.

You have to come to yourself. Every day witnesses an inexplicable rebirth, a resurrection of a self and its world no less miraculous than the resurrection of the flesh promised to those who believe in the next world.

Postscript

> Has someone started travelling
> towards our world,
> moving eyeless through the dark
> towards the light
> from nowhere to our place in space and time?
>
> Ray Tallis, 'Overdue'

In this chapter I have been circling around the distance between us and our heads. The impersonal processes that lead up to our head's creation and our insertion into historical and physical time that preceded us and will extend indefinitely beyond us are potent reminders of this distance. Another reminder, available as soon as we cup the skully frontal bone of our bowed head in our comforting hands, is that our head, the material object we supported on our shoulders, will outlast us.

In his wonderful memoir *Speak Memory* (1951), Vladimir Nabokov writes of 'a young chronophobiac' who 'experienced something like panic when he looked for the first time at home movies that had been taken a few weeks before his birth'. What was most disturbing was how little different the world seemed despite his non-existence and that 'nobody mourned his absence'.

What particularly frightened him 'was the sight of a brand-new baby carriage standing there on the porch, with the smug, encroaching air of a coffin'.[7]

It is a curious thought that one was once awaited by the world. That you were an expected baby, even though what was expected was defined in only general terms.

The empty pram is not the worst of it. The world enclosing you is but the minutest portion of the world without you. And yet this world-without-you, this 15,000 million-year-old universe, 100,000 trillion light years wide, populated by 6,000 million heads like yours, exists together, as a place for you to be or feel lost in, only in your head. It is your head that brings together things that exist, but do not coexist, to torment you with your own nullity.

chapter four
Airhead: Breathing and Its Variations

Now – for a breath I tarry
Nor yet disperse apart –
Take my hand quick and tell me,
What have you in your heart.

A. E. Housman, A Shropshire Lad, xxxii

Introduction

Our heads are endlessly trafficking with the atmosphere. From the first intaken breath, which quickly modulates into a howl, to the last gasp, there is a constant passage of air through the mouth and nose.

Life and breath go together. Life depends on the fact that the air that is taken in is richer in oxygen than the air that is breathed out and the air that is breathed out is richer in carbon dioxide than the air that is breathed in. How lucky, then, that breathing is so easy. Taking a breath involves lowering the diaphragm at the bottom of the chest cavity and moving the ribs out to the side, by this means increasing the internal volume of the chest. This sets up a pressure difference between the outside world – which begins at the mouth and nose – and the thoracic cavity. Air is driven down the pharynx, larynx, trachea, bronchi and bronchioles and unsqueezes the folded alveolar surfaces of the spongy lungs. The procedure is then reversed: the lungs collapse and air is squeezed out.

All very straightforward. Actually, no. It is not certain that, if it did not happen spontaneously, we would be able to 'do' breathing at all. Knowing how to move the diaphragm and knowing which way and how to move the ribs is not something that you could learn quickly enough to save yourself from death from lack of oxygen shortly after birth. Besides, breathing is something you have to keep up whatever other calls there are on your time, however preoccupied or busy you are, even when you are asleep or in a coma. Pity the poor knight in Friedrich de la Motte Fouqué's tale who was unfaithful to his fairy-wife, Undine. She cast a spell on him so that he had to carry out voluntarily those bodily functions which normally take care of themselves. There is a very rare disease in which children do not breathe spontaneously, due a developmental error in the part of the brain which regulates breathing. It is called 'Ondine's Curse'.

For the uncursed, breathing will happen, whether we will it or not. Indeed, we cannot stop it happening. Try closing down your airways. Within a minute or so, you are in a state of extreme discomfort, frantically gagging against your closed epiglottis. The relief! Those lovely lungfuls. Breath-holding attacks, by which children aim to terrify their parents, could never end in death: even if the child managed to overcome the impulse (that goes deeper and is more urgent than any other impulse) to breathe, until it lost consciousness, it would start breathing again as soon as coma supervened.

This link between life and breath was poignantly underlined as I sat by my father's bedside just after he had died. For a fortnight, we had listened to him hyperventilating as he resisted dying of pneumonia. His panting mined centuries out of hours. One afternoon, he seemed a little more peaceful and we awarded ourselves a break from our vigil. Hardly had we reached home than we were summoned back. He had died. The stillness and the silence as we approached his already cooling body were palpable: it seemed to radiate from his dead whiteness. I had never seen him entirely still; never seen him not in the grip of that incessant rhythmic movement that had accompanied him all of the days of his clenched and troubled life. The breathing and the troubles had come to an end.

How wonderfully this link between breath and life has played into our collective sense of the world. The *pneuma*, breath, *becomes* the spirit within us; and the respiratory tide, standing for the consciousness that animates our body, is projected into the idea of a spirit animating the world. The wind among the trees, driving the clouds across the sky, sending sheets of scales across the waters, is our soul writ large. As the writer Dudley Young has expressed it, 'the words for wind, soul, and breath commingle in virtually every language': 'Thus what moves the visible world-body, indeed pushes it around, is the invisible world-soul, which is wind, which is *pneuma*, which is divinity, which is God.'[1]

But what has this to do with the head? The head, after all, is only a convenient throughway to let air into the body. It is on its way elsewhere and the real business of air is done in the chest, in the lungs. A deep breath is a lungful not a headful; any exchange of gases in the head takes place via the blood which touches the outside air only in the lungs. Deep breathing does not clear the head or make the thoughts less stale.

Breathing is involuntary; but it may be subordinated to voluntary purposes. Breathing is unthinking; but it may be woven in with, shaped by, thought. The head is the place where this happens; where the human spirit, that collective consciousness that underpins the sense of the sacred and of the gods in among the things of the world, is outered or uttered in exhaled (and sometimes inhaled) air. For the head shamelessly appropriates the air that was not meant for it and suborns it to purposes many of which were unheard of before mankind. The lungs don't know what they are missing.

The most elaborated of these purposes are materialized in speech. It creates a new kind of connectedness supplanting the connectedness of the breeze animating entire forests, lakes, clouds: cities woven out of voices and voices weaving a new fabric of voices in the city. But there are other ways in which airhead expresses itself, the world, and its relation to the world. Let us deal with these other purposes now.

Normative Panting

'Man', Hazlitt said (citing Aristotle) 'is the only animal that laughs and weeps, for he is the only animal that is struck with the difference between what things are and what they ought to be'.[2] Nature is normless. (And, strictly *entre nous*, a bit gormless.) There are many ways in which 'is' and 'ought' part company and incongruity is registered.[3] We most commonly observe a gap between the one and the other when our expectations are overturned and it is precisely because we are creatures who entertain highly specific, explicit expectations, that we are always experiencing the unexpected. The unexpected is not, however, necessarily funny: we may be frightened, disappointed, or relieved.

Specifying the kinds of surprises that are laughing matters is very difficult indeed. What do custard pies in the face, a clever rhyme, a dirty story, sexual innuendo, jokes about blondes and Irishmen, sardonic footnotes, gurning and squinting, pretending to be drunk, catchphrases, assumed dialects or foreign accents, and tickling have in common? The answer is that nobody knows, though many have tried to pretend otherwise. Among them are numbered quite a few philosophers: Henri Bergson, Helmut Plessner for example. As Vic Gattrell has observed in his rich and wonderfully witty commentary on the social significance of laughter, if we examined the great range of modern laughter theories, we should 'expire of something other than laughter'. What is more,

> most have focused on specialized forms of laughter. None has tried to accommodate all laughter's repertoire in a single theory. This is understandable in view of laughter's many provocations. We can laugh at others in ridicule and derision, or from malice. We can laugh in triumph, joy, sympathy, surprise or satisfaction, or sardonically in misery or pain. We may laugh at incongruity or transgression, or at the verbal 'ambushes' of wit or pun.[4]

The huge amount of attention that laughter has attracted reflects a sneaking suspicion that it is connected with something that goes very deep in human nature. Just how deep was suggested by Nabokov:

> The beginning of reflexive consciousness in the brain of our remotest [hominid] ancestors must surely have coincided with the dawning of the sense of time…[and] the first creatures on earth to become aware of time were also the first creatures to smile…[5]

I like that link between smiling – joking and laughter – and the awareness of time. We are animals for whom time is not merely lived through the vicissitudes of our bodies but is experienced through an explicit past and an explicit future. Our time sense fits with our being creatures that entertain concrete and specific expectations. We have a feeling, on the basis of a recalled past, of how things are going to be in an anticipated future. And so we are aware of the difference between how things turn out and how they should have turned out. We are consequently susceptible to surprises: by punchlines that confound our expectations; by things that come together when they should be kept apart – as in rhymes and puns that neatly conjoin the disjuncted, or when sexual desire expresses itself in daily life. Just as comedy is so often based on the invasion of the body into the realm of the mind (walking into a lamp-post while showing off one's knowledge of Hegel), so puns, like rhyme, allow the material sounds of words to interfere with the realm of meaning, another instance of matter invading mind.

Laughter which registers the gap between 'is' and 'ought' also bridges it. Think of the laughter of embarrassment when we find ourselves uncertain how to behave, how to meet expectations, or how to cross the boundary between verbal and carnal intercourse. The multitudinous links between sex and laughter – laughing at others' sexual desires and behaviour, lovers laughing away the grossness of the standard expressions of sexual desire and the social and physical awkwardness that comes in its wake and so on – would fill an entire volume.

If funny ha-ha seems funny peculiar, it seems even more so when we switch our attention from the theory of laughter to the phenomenon itself. Let us eavesdrop on a group of people who have got together 'to have a laugh'. We are reminded of what experts tells us: that laughter is rarely a response to set-piece jokes and stories: 'Mutual playfulness, in-group feeling and positive emotional tones – not comedy – mark the social settings of most naturally occurring laughter.'[6] In short, banter. We may have got together for a laugh but we seem to be laughing as a way of asserting our togetherness. Not to share the laughter is a refusal to join in. 'I do not find that funny' says 'I do not share your picture of the world. I do not accept you.' When the laughter is almost continuous, and it often is, for laughter breeds laughter, you can hear how strange it is.

The dialects of laughter seem almost infinite. Sniggering, hooting, braying, giggling, bellowing, tittering, chortling, chuckling, all have their own sonic structure. What is more, they have their distinctive occasions and settings. There is a laughter code as precisely prescribed as a dress code. The manner of your laughter contributes to the impression others have of the kind of person you are. The paradigm gigglers are nervous, impressionable schoolgirls, who may also titter (covering their mouths partially to mute any vocalization that might escape) with a tinge of malice. (In the mouth of a master comedian such as Frankie Howerd, the word 'tittering' can make titterers of us all.) Upper-class males bray, as they bond with their kind and pronounce from wealthy, sporty heights upon the unwashed world around them. Sniggerers are furtive and mean-spirited, in keeping with the word itself that is within calling distance of 'snivelling' in the dictionary, and somehow reminds us of the snot that lies in the nose of those whose laughter is close-mouthed, so that the necessary expiration is diverted through those mucky passages. Chucklers and chortlers are altogether nicer – less directed against others, more purely joyful. There is a kind of warm sunlight in their laughter and they probably sometimes rub their hands with innocent Dickensian pleasure. Bellowers are attractively gripped by their laughter, so long as they do not go on too long. At any rate, bellowers seem reasonably good-hearted. The belly laugh may seem more genuine and less malicious than the brain laugh which is very pursed up and kept to itself. Hooters dole their laughter out in great blobs their hooters could not accommodate.

There are, of course, modes of laughter that are provoked by pain – the wild and hellish laughter of the damned, the hysterical laughter of the panic-stricken – whose sonic structure may give less of a clue to meaning than the circumstances in which they are laughing and the accompanying facial impressions. The phrase 'sonic structure' seems rather po-faced for something as spontaneous (or originally, at any rate – we all know how to laugh when the circumstances require it)[7] as anarchic as laughter.

The casual appearance, however, is deceptive. Laughter's *déshabillé* conceals a tight corseting of rules governing its acoustic structure. This has been studied *con amore* by that student of matters gelastic, psychologist and neuroscientist Robert R. Provine. The basic action is pretty straightforward. On the basis of sound spectrographic studies of giggles, shrieks and belly laughs, he found that laughter had its 'distinct signature'.[8] A laugh is composed of 'a series of short vowel-like notes (syllables) each about 75 milliseconds long, that are repeated at regular intervals about 210 milliseconds apart'. This highly disciplined sonic structure is subjected to a further discipline. A given laugh has to stick with a particular vowel sound: laughers may choose 'ha-ha-ha' or 'ho-ho-ho' but not 'ha-ho-ha-ho'. There is resistance to producing such acoustic mongrels. Where there is variation, this usually involves the first or the last note in the sequence. Provine gives 'cha-ha-ha' or 'ha-ho-ho' as admissible variants.

This structure is evident in all forms of laughter. That is why we can recognize that guffawing and giggling, hooting and tittering all belong to the same behavioural family. There are a couple of other features. The explosively voiced blasts of a laugh have a strong harmonic structure, with each harmonic being a multiple of a low (fundamental) frequency – about 502 Hz in females and 276 Hz in males. And the laugh is temporally symmetrical: a bout of recorded laughter played backwards sounds similar to one played in the direction in which it was produced. The only asymmetry is in amplitude: laughter tends to decrescendo as we move from the earlier to the later notes, as the air supply runs out.

I have dwelt on the sonic structure of laughter to emphasize its almost ludicrous strangeness, highlighted when we think of the range of its objects. While (I suppose) we could imagine some connection between tickling and stereotyped, ordered panting, it is difficult to think of the link between this panting and a sardonic observation about a mistranslation of a passage in Aristotle without breaking into another instance of normative breathing. The mismatch between the frequently sophisticated or at least abstract objects of laughter and the process itself is a poignant testimony to our peculiar condition; to the way in which we have to use the givens of our organic body to express, interact with, all sorts of contingencies and circumstances that our bodies could not have foreseen. It is not as strange of course as in sex – that ubiquitous source of laughter – where the plunging of an erect penis into a self-lubricated vagina has to pass itself off as the supreme and appropriate expression of emotions awoken by the hope of being valued for one's self, or by sonnets, while at the same time expressing a delicious transgression of the world from which these things originate.[9]

There is nothing straightforward even about the connection between laughter and tickling. And the fact that one cannot tickle oneself casts some very interesting sidelights on what it is to be embodied, and the relationship between identification with one's body and experiencing that body. As the neuropsychologist Chris Frith and others have found, the reason we cannot tickle ourselves is that the brain normally blocks sensations caused by the body's own movements, and these include the movements associated with self-tickling.[10] Self-inflicted sensations are predictable: the motor commands, which circulate around the brain, tell the brain what to expect. Since the brain's job is to deal with the unexpected, the entirely *expected* sensations associated with self-tickling are scarcely registered. When, on the other hand, we are being tickled by others, we do not have the imprint of the motor activity circulating around our brain telling us what to expect. The resultant sensations, unpredicted and unexpected, are not therefore suppressed: they are experienced to the full. Self-tickling via a capricious robot that transmits the tickle at unpredictable intervals is like being tickled by another: very ticklish, confirming that ticklish sensations are those that do not map onto a motor programme and cannot be predicted. And the threat of tickles from others activates the brain in the same way as actual tickling: we predict the unpredictable.

Tickling is important for another reason. It appears that this is a stimulus to laughter in non-human primates, most notably chimpanzees. Actually, theirs is not really laughter. Tickled chimpanzees, according to Provine, produce a 'breathy panting vocalization' during each inspiration and expiration. By contrast, in human laughter a single inspiration is segmented into a succession of sharp-edged vowel notes. What is more, the vocalizations in chimpanzees are directly plugged into the here and now, lasting only as long as the physical stimuli that provoke them: tickles, play etc. When the tickling stops, so does the panting. Nothing could be further from the paralinguistic phenomenon of human laughter with its exotic, often higher order, frequently abstract, referents; nor from that sense of the discrepancy between how things are and how they ought to be that is the most fundamental provocation to human mirth.

To acquire a sense of this disjunction, one has to have a fully developed sense of a state of affairs, 'out there', independent of one's own experiences. A sense, that is, that something or other is the case. This 'Propositional Awareness' is unique to humans for reasons that go right to the heart of human nature.[11] Only when we have a fully developed sense that things are the case, can we compare them with how they might have been. This comparison takes place only in human heads. That is why my head and heads like it are the sole objects in the universe that find other objects in the universe funny and laugh at them.

Laughter itself is laughable. A person who collapses into giggles which prevent her from telling the funny story coerces us into finding the story, when it comes, to be funny. There are even jokes about people failing to be funny. This can become very enfolded indeed, as in the late comedian Bob Monkhouse's joke about himself – or not about himself: 'They laughed when I said I wanted to be a comedian. Well, I became a comedian. They're not laughing now.'

Laughter is infectious. We are inclined to laugh (at first at any rate) when we are in the presence of someone else who is laughing or even when we listen to canned laughter. Just how infectious is illustrated by an episode that occurred in 1962 in a girls' boarding school, where, for a period of six weeks, the school was forced to close.[12] It began with three girls who were stricken with bouts of uncontrolled laughter for hours on end. Eventually nearly half of the 159 school boarders were affected, laughing for up to sixteen days at a time. The school was closed and the children sent home but this resulted in further spread of the condition throughout communities and to other schools. Attempts to reopen the school were disastrous. In the two-year period that the gelastic epidemic lasted, fourteen boarding schools and entire villages and towns were affected round the eastern shore of Lake Victoria. In a place called Nshamba in Tanzania, attacks of uncontrollable laughter affected more than 200 people in a population of 10,000. While nobody died, there was much agitation, embarrassment, exhaustion and interference with daily life.

Laughter under such circumstances is no laughing matter. And there are other occasions when normative breathing may be extremely unfunny. While laughter is uniquely human, it may also be uniquely inhuman. When we laugh together, we affirm the norms, and our togetherness. And so we laugh together at those who do not know the codes, who are different from us. Laughter is one of the most potent weapons of the bully and the mob. Sneers, jeers, jibes, mockery that aim to demean and humiliate, are the staple of everyday cruelty. The laughter that accompanies the insult, or the damning judgement, validates what is asserted, but at a level that goes deeper than justice, than logic. What is more, it recruits the judgement of others. The infectiousness of laughter translates into infectiousness of judgement: your laughing judges turn to each other and find confirmation in each other's laughing faces of the judgement they are passing. United by laughter, they amount to a formidable team, a tower of ridicule, that dwarfs the laughed at. Against such laughter, which corners the one in an unmediated judgement passed by the many, there is no appeal.

Dealing with ridicule requires grace, or at least fancy footwork. The good sport joins in the laughter, pretending that the hurtful and the humiliating is excellent fun. But the bully knows how to exploit even this – how to cross the line between gentle, affectionate teasing and malignant mockery – and how to push things past the point where they can be shrugged off. When his victim finally takes offence, he will then accuse him of lacking a sense of humour. After all, the lack of a sense of humour is grounds for more laughter and the failure to get or take a joke is itself a joke, particularly if the one who fails is a standing butt of jokes. As Celia said in *As You Like It*, 'the dullness of the fool is the whetstone of the wits'.[13] There will be some who spend much of their lives bathed in a vile mixture of their own sweat and others' glee.

Humans are not only the sole creatures who laugh; they are also the only creatures who reflect on laughter. They plot ways of bringing it about, think about what is funny and what is not, wonder (sometimes ruefully) why some people find some things funny and others do not, and acquire a livelihood through an entertainment industry devoted above all to sex and laughter, and often the two together. They even write laughter down, formalizing those vocalizations as 'ha-ha' and 'ho-ho' and 'hee-hee' and 'tee-hee', passing judgement on the laughter and the one who laughs by representing it one way rather than another. I would rather be written down as one who 'ho-hos' than as one who 'tee-hees' which seems rather too close to the glistening yellow of malicious glee.

The most destructive laughter exploits the risk of exposure that comes from acting on sexual desire. The laughter that greeted Alexander Pope's request for his friend Lady Mary Wortley Montagu's hand in marriage – believing perhaps that his genius would compensate for his four-foot-six-inch stature (he was described by his enemies as 'the hunch-backed toad') – still conveys undiminished horror a quarter of a millennium after the laugher and the laughed at are both dead. And while Camille Paglia's claim that Medusa's 'hideous grimace' represents 'men's fear of the laughter of women' seems insufficiently substantiated, it does have a profound psychological plausibility.[14]

The saddest laughter is the laughter of the mad. Inappropriate laughter is an early sign of psychosis. The person with incipient schizophrenia may laugh at personal tragedies, or the tragedies of others – perhaps as a way of coping. And we are all familiar with the eccentric person who laughs mainly to himself. His laughter is not a mode of communication, a shared and sharing response to something that is awry in the world which he copes with through solidarity with others. Instead, his laughter marks the distance between himself and the world of others: they are how things are; within him is how things ought to be. Under normal circumstances, we laugh thirty times more frequently when we are with others than when we are alone; for the madman this is reversed. He does not laugh with the world but at it; and the world rewards this by laughing at him.

Laughter ultimately is a marker of our being eccentric creatures, as Helmut Plessner said, sticking out in the world in which we are incompletely dissolved, incompletely at home. We laugh sometimes so that we shall not cry. And sometimes we laugh until we cry 'tears of laughter'.

Some Other Uses of Intercepted Wind: Coughing, Yawning, Sneezing

For a structure that has only a (literally) passing interest in air, the head seems to do an impressive number of things with it. Coughing, yawning, sneezing, snorting, harrumphing and simulating orgasm, indicate the extraordinary diversity of ways in which head breezes can be utilized and the range of tools that can be fashioned out of air on the way to, or on the way back from the lungs.

We fit very well into the natural world in which we first found our fortune before we created a parallel human world that almost occluded nature. The excellence of fit, however, is only of a general kind. It overlooks the variations of circumstance, the contingencies of particular places and particular times. Consequently, even though breathing in general is a good idea beautifully executed, there is the ever-present possibility of glitches. When, for example, I take a deep breath, I might entrap not only the air I need but one or two things I most certainly do not need: dust, samples of the organic world such as flies or other biological interlopers, fragments of food or other artefacts. If these were not expelled, it would not take very long before the lung cavities were infilled and they became as solid and useless as a pair of rocks in the chest.

Hence *coughing*: a short intake of breath, followed by a temporary closure of the larynx, followed in turn by contraction of the muscles of expiration, and an opening of the larynx. Under the high-pressure blast of air thus created, the airways are cleared of dust, aphids, biscuit crumbs, or home-brews such as phlegm. The cough reflex is triggered by irritation of the sensory nerves in the lining of the airways. This irritation is experienced as an intense, unbearable tickle.

Coughing is almost irresistible. Even the thought that one might cough prompts the beginning of a desire to clear one's passages. This thought is particularly compelling when it is important that one should not cough: in the soul-freezing quiet passages of a piano concerto one loud cough can annoy 3,000 people; or when one wishes to observe another while remaining unobserved oneself. Anne, the baby of Enid Blyton's Famous Five, was almost defined by her overwhelming desire to cough when they were spying on crooks engaged in self-incriminating acts. (Our sympathy for Anne stitched another small thread in the fabric of human solidarity.)

Nothing, therefore, could be more biological and unwilled a mechanism than coughing. And yet, humans being what they are, this event with a clear physiological purpose, an unequivocal meaning, is ripe to be transformed into a human symbol with a multitude of sometimes profoundly ambiguous meanings. Coughs, unlike shrieks and cries and calls, are only incidentally noisy; their sound has nothing to do with their function. The blast of air, if somehow muted, would still expel the trespassing material. It is we who expropriate the noise of the cough and make it serve our own purposes. We manufacture a cough in order to have the noise of a cough. We do this to draw attention to ourselves; to say 'Here I am!'; or simply 'I!' The cough, in short, is coughed in order to make evident the presence of the cougher.

This use of the cough, prompted not by a tickle but by a desire to makes one's presence doubly present, grows out of a rather sophisticated sense of one's own being, one's being for others. Just how sophisticated is made clear by further developments. The cough is made to stand for the person who feels he or she has been overlooked. Such a cough has many variants. One of the most sophisticated is 'the modest cough of the minor poet'. This cough, or the idea of it, draws on, or reaches into, several tussive dimensions: the throat clearing that may precede a public recitation; the coughing that may signify nervousness; and the aforementioned pointing to one's own presence. Throat-clearing, preparatory defrogging of the throat, may itself be isolated as a marker of the public speaker, or the speaker on the verge of saying something rather difficult (preparing the voice box and dealing pre-emptively with nervousness). This in turn may be stylized, as when we cough in inverted commas to command silence or utter the written word that captures a certain highly formalized cough: 'Ahem!'

Such sophistication doesn't stop coughing being an annoying currency in the economy of everyday life. The cougher – voluntarily or involuntarily – says: 'I am hurting', 'I am ill', and 'I might choke', 'I am in need'. Children soon learn to assert the power of the powerless through this sound that harpoons the attention of anxious parents. It may operate in the other direction. I once worked with a technician, an unmarried only child, whose entire life was ruled by her elderly mother's 'little tiddly cough'. The cough had caused holidays to be cancelled and at least one disengagement from a long-standing boyfriend. Full marks to the cognitive behavioural therapist who gave her the inner strength and confidence to overcome the maternal 'tussocracy'. She married and has children who, I like to think, have kept their grandma awake with *their* 'little tiddly coughs'.

Tiddly coughers exploit folk memories of more serious symptoms. That folk memory, particularly associated with tuberculosis, has been upgraded by AIDS from memory to omnipresent reality in many places in the world, where the passage from coughin' to coffin is too well trodden to be appreciated as a pun.

The minute lay and professional attention that coughing commands has spawned a rich typology: fruity and dry, hacking and barking, and so on. And beyond pathology, there are other types of cough: the warning cough and even the cheating cough. Readers may recall the infamous episode in the quiz show *Who Wants to be a Millionaire?* when the contestant was able to draw upon a supplementary knowledge bank in the head of a confederate in the audience who coughed to indicate the right answer.

Lecturers famously pull off the extraordinary feat of talking in other people's sleep. The cardinal sign of an audience undergoing inner emigration is the *yawn*. Experienced academics are familiar with the more striking features of the yawn, such as the mouth open so wide that you can see the student's oesophagus without the aid of an endoscope, or the three-note expiratory phase sounding like a lost fragment of Palestrina. Equally familiar is the suppressed yawn of more polite students, aching with tiredness at having risen 'at the crack of yawn' but mindful of their future and hence of your judgement upon them: the flaring of the nostrils, the bellying of the under-jaw, the subtle warping of the face, the stray tear.

Unfortunately for lecturers, yawning is infectious like laughter, only more so. Stamp collectors may like to know the meta-yawning fact that 50 per cent of people will yawn within five minutes of seeing someone else yawn. Even thinking about yawning is enough. It is a safe bet that, while you might have read through the earlier section on laughter entirely po-faced, you will have yawned by the time you have reached this sentence. Yawning is also like laughter in respect of its mechanics, though it is somewhat simpler: a deep inspiration and shorter expiration are separated by a brief, copestone pause. This may be subjected to embellishments; yawn-spotters will observe the grunts, cries, notes and other acoustic curlicues that grace the simple arch.

Surface similarities highlight the profound difference between laughter and yawning. At its best, mirth records joyful surprise at the variousness and rumpled unexpectedness of the world; at its most typical, yawning responds to the sense that the world is monotonous, ploddingly predictable, joyless, flat. The cliché, the recycled joke, the pitiless detailing of facts, unvarying prospects, endlessly repeated tasks – this is what fuels the sense of tiredness that triggers up the yawning centres. Laughter says that the world is astonishing, yawning that it is obvious, exactly matching expectation.

And yet the obvious is not itself obvious. Philosophers, whose job requires them to untake the for granted, have often been inspired by the mystery of boredom. Martin Heidegger, perhaps the greatest philosopher of the twentieth century, 'proposed to demonstrate the birth of philosophy from the nothingness of boredom' and reflected on the strange fact that we can find this complex, rich, infinitely various world, and our equally complex and various lives in it, more than a teeny weeny bit dull.[15]

It is fitting, therefore, that yawning – which, it has been estimated, we do about 250,000 times during our journey from the cradle to the grave – should resist easy understanding. The sheer number of explanations is a clear signal that it lies beyond current explanation. It is essentially a large intake of breath. This points to a possible reason for yawning: we breathe less when we are drowsy, so that oxygen levels fall. Yawning is a response to the need to restore blood oxygen. Unfortunately, breathing oxygen when we are drowsy does not suppress yawning. Collapse of theory.

Other theories have focused on the infectiousness of yawns. Ape-watchers have observed that when one ape yawns, its yawn – the alpha-yawn, presumably – not only sets the others yawning but also seems to prompt movement to another location and engagement in different modes of behaviour. Perhaps yawning is a means of communicating and coordinating social behaviour, a kind of nod of agreement: 'This sucks. Let's go'. There are many problems with this explanation. First, the propagation of yawns is somewhat random and, even in apes, the association between a good-going outbreak of yawning and change of location and behaviour is somewhat weak and evidence for a causal relationship even less impressive. Secondly, reading across to us from apes (who have an entirely different world picture) is always hazardous. What is more, all vertebrates yawn and reading across from blue tits and crocodiles to us would seem to require even more unprotected leaps of the imagination. Thirdly, humans yawn quite frequently when they are alone and it would be a mark of desperation indeed to argue that the solitary yawner is communicating with virtual others. Fourthly, we know that human foetuses yawn from about the eleventh week onwards. They have no one to signal to, as their confinement is usually solitary. It seems unlikely that they are anticipating the dull lectures the extra-uterine world will offer them. Equally unlikely is the notion that they are bored with their intra-uterine lot. Although nine months of solitude with only muffled rumours for a world might be a bit tedious for such as ourselves, one might expect that the foetus would find growing from a speck of living matter to a seven-pound infant with a strong sense of needs and rights would be engrossing enough.

Neuroimaging, which demonstrates the activation of different parts of the brain, yields some interesting facts about yawning. When subjects are asked to observe videotapes of people yawning, parts of the brain light up that are usually associated with responding to social cues. They don't light up in the same way when the subjects are asked to view random mouth movements. The bits of the brain that light up, however, are not those associated with the extensive mirror-neuron system. This system is activated when we want to imitate an action we are observing.[16] From this, scientists conclude that contagious yawns are automatic rather than truly imitated actions that would require some understanding. We need to treat this conclusion with a little caution, as we shall see.

The simplest explanation of yawning, which strikes me as the most plausible, is that it is a *protective lung reflex* that maintains proper lung inflation and, in particular, prevents the bubbles in the lung sponge – the so-called alveoli where the exchange of oxygen and carbon dioxide takes place – from collapsing. In this respect, it is a little like coughing but, as with coughing, we have to acknowledge that no traffic into or out of a human head is, or remains, simple. Unlike non-human yawners, we deliberately use our yawns to signify those things that typically trigger involuntary yawns. We not only yawn with our fellows, we also yawn at them.

To be yawned at is as wounding as to be laughed at; to be found boring even more dismaying than to be bored. This critique of ourselves is especially damning because it is seemingly involuntary, and consequently sincere. As a signal of boredom, it can of course be deliberately elaborated. When we yawn in company, politeness demands that we should cover our mouths so that others are spared our exhaled air and a ringside seat view of the inside of our oral cavity. In order that the biological business can be transacted, coverage has to be intermittent. The polite yawner therefore pats rather than seals his mouth. The action in turn can be used to symbolize a yawn and, through this, the yawned at. An otherwise unassailable authority can be subverted; an evening out, a holiday abroad, a job, a marriage can be summarized, and judged, simply by the action of patting one's open mouth.

This manner of communicating our superiority to the world in which we find ourselves and which we take largely for granted finds us a long way from the automatisms observed by neuroscientists, far from disaffected apes, and far from the business of upholding the exquisite architecture of the lungs or maintaining the blood oxygen levels. No yawning matter, I would venture.

Sneezing is rarer than coughing and performs fewer communicative roles. We do not, typically, draw attention to ourselves by sneezing. Its biological purpose, however, is rather similar. The trigger in this case is irritation of the mucous membranes of the nose or throat – by dust, pollen, invading viruses, snuff, etc. The mechanism is a bit more complicated. The nerve endings in the nose are stimulated and impulses pass up the trigeminal ganglion to a set of neurons in the brainstem collectively termed, wait for it, 'the sneeze neurons'. These send impulses along the facial nerve back to the nasal passages, causing them to secrete mucus, and sometimes also to the lachrymal glands to squeeze out a few tears, and to the muscles around the eyes to cause them to close. At the same time, the sneezing centre sends impulses to the respiratory muscles via the spinal cord, resulting in a forceful expiration. By this means is generated both the vehicle used to wash out the offending material and the respiratory explosion to send it cannonading on to a handkerchief (in last century or so) or the back of the hand (preceding millions of years).

The slightest recollection, the most casual research, opens up bottomless drawers. Dig a little and you are in free fall. The next time you call out 'bless you' to a sneezer, remember what a venerable history it has.[17] The salutation is a link in a great chain or network of benedictions, connecting Tiberius Caesar (a melancholy unsociable man), who 'exacted a benediction from his attendant whenever he sneezed'; the black death, linking terror and sternutation; Icelandic invocations of the help of God; nineteenth-century Equatorial Africa, where, according to the explorer John Hanning Speke, the only trace of religion among the natives was 'the practice of uttering an Arabic ejaculation or prayer whenever a person sneezed'; and Thailand, where the Supreme Judge of the world is always turning over the book containing the life and deeds of every human being and, when he comes to your page, you will sneeze.

There is a famous moment in Xenophon's account of the

campaign of the Greeks against the Persians – *The Anabasis* or The Way Up. Xenophon, an Athenian general, gave an uplifting oration to his soldiers to follow him to liberty or death. At the end of his oration, somebody sneezed, a morale-boosting sign of the support of the gods. However, Napoleon's sneezing during the early hours of the morning before the hideous slaughter of the Battle of Borodino brought few blessings: 80,000 men were killed, uncountable others were wounded and general ruin was spread over Russia and, ultimately, France.[18] A less harrowing instance of a sternutatory interaction between nations is the story of a rather patrician English girl sneezing on a bus in Germany. The middle-aged German man behind her offered her the standard *Gesundheit!* She turned around in delight and said 'Ah, you speak English!'.

For reasons that are not exactly clear, sneezing can be quite pleasurable. (No one complains of a 'hacking' sneeze). Indeed, to be on the edge of a sneeze that does not arrive is highly frustrating: it is like being denied a micro-orgasm. The 25 per cent of us whose sneezes may be photically induced hasten to look at the sun or at a strong lamp to turn an incipient sneeze into the full blast before the feeling passes. Snuff-takers make sneezing into a major recreational activity, blackening their nostrils and their handkerchiefs and the little space between the tendons at the base of the thumb ('the anatomical snuff box') with tobacco dust in pursuit of minor ecstasies in which the body momentarily possesses its owner.

This brings us to our penultimate example of (largely) non-verbal breathing: that associated with the *orgasm*. More particularly – because it is associated with breathing – the female orgasm. More particularly still – because it illustrates the general principle to which we have just alluded – the faking of same. The most striking, or public, aspect of the female orgasm is the panting, whimpering, sobbing, cries of astonishment and delight, that accompany or signal it. While orgasms don't have to be noisy (sighs aren't everything), the vocalizations seem to be evidence that the male lover is a good lover, that he is loved for his love-making, that he has privileged access to the soul of the woman, that he is 'hitting the spot'. The panting, therefore, signals a multitude of things, but none of them unequivocally. They include gratitude for the experience and praise for the performance that brought it about. Beyond this, it says: you have taken me to another place, to a special place where we have enjoyed a very special togetherness. Or it may signal none of these things. The most extreme sexual intimacy can still clothe itself in ambiguity.

And so, in the end, to something less delicate, and perhaps occupying more human time: grunting. The topic is so large, I shall narrow it down to the disapproving grunt, which may attract its own measure of disapproval, and more specifically, to the harrumph.

Harrumphing is worthy of our attention if only because of the discrepancy between the seeming crudeness and unvarnished physicality of its means and the abstraction and sophistication of its occasion. A harrumph, often close to a suppressed bark, is an acoustic blob, a protolinguistic Ur-phoneme. This notwithstanding, it is typically triggered by rather advanced stimuli – for example items in broadsheet newspapers revealing some new fashion or trend in that notional object, that boundless rumour, called 'the world'. Harrumphs are particularly associated with the idea of a member of the establishment, whose overweight body provides the perfect instrument for manufacturing it. That body is likely to suffer from high blood pressure, due to a chronic state of being at odds with much that happens in the world around it, as well as the aforementioned obesity, one pertinent local manifestation of which will be jowls that shake when the expiratory 'gruff' subsides into a sideways titubation of negation. The body may well have been educated in a public school, and even had a period in the army, will be resistant to the new, if only because change threatens its privileges, and could be clad in a dog-tooth check jacket. It will hail from the shires, though its harrumphing typically takes place indoors and probably an urban indoors, indeed a gentleman's club. There, surrounded by the like-minded, it will be reinforced in its harrumphing tendencies as its harrumphs are rewarded by echoing harrumphs. It may precede the angry turning of the page of a newspaper, where new harrumph-worthy outrages will be revealed. As the preceding paragraph has shown, the ease with which we manufacture and recognize stereotypes is shared by har-rumphers and harrumphees alike.

Winding Up

We have by no means exhausted the repertoire of headwinds. Think of mming, humming, umming and ahing; of gasping (with surprise or pain); of sighing, snorting and snoring. Think of whistling that has so many different functions: entertaining or comforting one's self; populating the dark with one's favourite tunes to make one fear its inhabitants the less and, by magic think-ing, tame them; signalling one's presence or surprise; and sexual harassment. Or of that end-stopped slice-of-inspiration the hiccup, for so long the iconic sign of the staple of comedians, the drunk; and tut-tutting – admittedly more meaty than airy. Yodelling, which migrated briefly from its native grounds to the United Kingdom in the early 1960s of the last century. And we must not forget spluttering – when biscuits have gone down the wrong way; or speech has collapsed under the burden of the emotion it is trying to convey; or an actor wishes to signify shock or surprise. On top of all these pure strains, there are marvellous hybrids or com-pounds: the laugh-cough-gasp-splutter mobilized to indicate stifled involuntary amusement and amazement; the smile-yawn seasoned with exhaled cigarette smoke; and tut-tutting accompanied by anaerobic frowns and windless head-shaking. And sequences of headwinds – as when a yawn causes laughter or an ill-judged laugh prompts sharp intakes of startled and disapproving breaths. Not to speak of all those assisted, transformed, instrumented expirations that are trumpeted, tromboned, fluted, and so on.

The list is endless but I want to draw this chapter to a close by briefly reflecting on how far we have travelled from the anatomical structure we looked at in the mirror and from the physiological master chef producing the impressive range of secretions we examined in chapter 2.

In that chapter, we got some idea of what we headed ones do with the material that our heads serve up to us, with the saliva and sweat and mucus and tears and so on, that seep out of our heads for a variety of purposes. The ingenious way we subordinate these no less ingenious physiological secretions to non-physiological, symbolic human purposes, the way they are woven into human culture and history, was striking evidence of our distance from our own heads. The head transcends itself.

We have seen from this chapter that it is possible to subject even simpler materials, the air entering and leaving our body, minding its own physiological business, to even greater transformations, to make human actions out of biological events. And this is only the beginning. For we are on the edge of a great space of possibility inflated by the most powerful of all headwinds: speech.

chapter five

Communicating With Air

This is the epoch of the vowels; before that it was the
sound of the crickets; and before that the sound of the
wind in the trees.

Gottfried Benn, *Primal Vision*

This mighty wind, building up over 100,000 years to a gale created
by 6,000 million heads, has blown open a space of possibility in
which our individual and collective consciousnesses have their
being. I approach our theme with foreknowledge of defeat. Speech
about speech is problematic. How can speech speak itself? But the
question answers itself: speech contains the word 'speech' which
encompasses all speech; and there is the word 'sentence' which signifies all sentences, and the word 'word' which encompasses all
words. We therefore have the illusion of a verbal viewpoint on all
verbal viewpoints. We have words that seem to gather together all
the pulling apart and gathering together that is speech. And it is
one of the most fundamental and revealing truths of human language that it is riddled through and through with metalanguages –
language about language. Speech, the home of the most exquisitely elaborated self-consciousness, is conscious of itself. When I
speak, I know that of which I speak, why I speak, to whom I am
speaking, what I hope to achieve through speaking, and that I
am speaking.

We have entered dizzying territory. Let us note, simply, that while we seem to be able to speak speech, to talk about this billion-stranded conversation humanity has had with itself, what follows is an attempt to capture the ocean in a child's bucket itself made of a bit of the ocean.

Speech is fashioned out of exhaled air, exquisitely sculptured by the lips, the tongue, the palate, the throat, working together in marvellously rapid synergy, to fashion plosives and labials, the stopped and the unstopped, open vowels and closed consonants. These sounds are linked into utterances of bottomless complexity, in which we at once discern meaning and intent. Their rising and falling tones and volumes declare that what is being uttered is an assertion or a question, or a wish, or a sigh, or a curse; that it is gentle or aggressive, flattering or mocking, informative, misleading; that it wants not only to be heard or also overheard. As it is spoken, it can mock, parody, itself; it can protest its own sincerity, make an earnest of its earnestness.

Thus the breath that is life articulates the consciousness that life sustains. We already marvelled at how our ancestors intuited a link between the world within and the world without, the human spirit and the natural world, in the *pneuma*, the God whose power was evinced in the wind that animates the trees, making the woods move as one; linking the land with the sea, the ruffled feathers of the birds with the scudding clouds and the booming caves. The *pneuma* that is in speech gathers up the collective human consciousness that is human nature, the natural world refracted through our communal understanding, the shared human world, the individual worlds to which our individual selves relate. This second nature reaches its most elaborate expression in the city, that network of voices, each connecting up so much of the shared out there, each taking its place in a boundless nexus of speakers. The cacophony of cities is humanity's answer to the winds that preceded mankind, to the storms that broadcast nature's indiffer-ence to its latest child, the one in whom and by whom it is known.

Like thermals that make distant mountains tremble, so our thoughts, made outward and audible in puffs of air, have changed the face of the earth – by that activity that is often dismissed with the literally accurate term of 'gassing'.

Speech, which uses abstraction, says how things are, denies how they are, says how they might or ought to be and, most funda-mentally, says *that* they are. It accompanies us from morning to night, from the end of infancy to the end of dotage. And when we are not speaking out loud, we are speaking to ourselves in the windless speech of thought. We talk ourselves into and out of emotions – fear, hope, anticipation – into and out of plans, even plans for talk. We instruct, encourage and guide ourselves. We even speak in dreams and dream that we are speaking.

Our heads are able to be so speech-bound because our mouths are not busy grazing, catching dinner, or weaving nests. They are free to fashion this exquisitely sculptured head-zephyr. It requires the coordinated activity

> of a bellows-like respiratory activator, which furnishes the driving energy in the form of an airstream; a phonating sound generator in the larynx (low in the throat to translate the energy); a sound-moulding resonator in the pharynx (higher in the throat) where the individual voice pattern is shaped; and a speech-forming articulator in the oral cavity (mouth).[1]

This seems rather a lot to expect of a two year old and yet there are few two year olds who cannot verbally indicate their basic needs – or, more precisely, their wants.

What is surprising, then, is not that no other animal speaks but that we do. And we haven't been speaking for very long – perhaps

40,000 years, since the time of Cro-Magnon man.[2] So speech is a very late comer. We have no idea by what means it came. The earliest theory was that it came from the gods, which is just what you might expect humans to think. Speech, which embodies collective consciousness, is greater than any one of us. As the sociologist Emile Durkheim argued, we project this sense of society, into the idea of a sacred Being. Nowadays we prefer to think that the gods came from language: in the beginning was the Word, and the Word made God who preceded the beginning. We opt for naturalistic, rather than supernatural, accounts of the birth of speech. Such theories, however, tend to be Just-so stories, which many linguists treat with a contempt reflected in the names given to the theories.

The 'ding-dong' theory has it that speech arose out of imitation of the sounds of things – which makes it difficult to account for the words we have for silent things, and abstract entities. This limitation is even more obvious in the 'bow-wow' theory according to which the earliest speakers imitated the sounds made by animals. Neither of these theories, what is more, accounts for the precise meaning of words; for the grammar which dictates how they can be combined in different ways; or for the fact that they may be uttered in different tones of voice, to mean different things, or to participate in different kinds of speech acts – questions, orders, curses, greetings, and so on. This criticism applies with equal force to the seemingly more promising notion that language grew out of sighs of pleasure and moans of pain ('pooh-pooh' theory); out of oral transcription of hand gestures ('ta-ta' theory), from work chants ('yo-he-ho' theory), or from warning grunts ('uh-oh' theory).

The fact that we can seem to reconstruct the transformation of expiration to information by starting from so many different places should warn us against any of them. Nevertheless, they cast a prospective light on speech itself: we see an aspect of what speech is when we link it with a putative origin. Any explanation of language, of the most obvious way in which we make things explicit to ourselves and to others, must account for the fact that, more than anything, we are *explicit animals*. The propensity to transform expired air into sounds that are used to refer to other things, to possibilities that may or may not be realized, presupposes an explicit sense of oneself, of the material world, and of other humans. The origin of that higher level of self-consciousness and of consciousness of other things and other people is a profound mystery.[3]

Supposing speech really did grow out of laughter, as Robert Provine has suggested, then this would remind us of two things.[4] First, that speech is highly disciplined, segmented, organized breathing. Laughter, too, has its disciplines, as we discovered earlier. Unlike the see-saw panting of the tickled chimp, the laughter of humans is broken into repeated vowels, emitted at fixed intervals and with a controlled duration. There is almost a phonology of laughter, a control of the choice of sounds; and almost a grammar of laughter regulating the choice of successive sounds. Those who are committed to 'ho-ho' cannot interpolate 'ha' or 'hee'. And laughter, which registers the difference between how things are and how they ought to be, which says 'just fancy that!', is almost referential, noting *that* such and such is the case. There are, of course, profound differences. The sounds of language belong to a *system* of sounds. The phonemes sound as they do only by contrast with other phonemes: a unit of verbal sound realizes a place defined by its difference from other sounds. And the grammar of language has a complexity that has defied characterization in a finite set of rules.

Even so, the gelastic theory touches on something central to language, and which laughter by virtue of its deficiencies makes evident. Though it may be used for cursing, for threatening, for carrying out an endless variety of speech acts whose primary aim is not to inform but to operate upon the consciousness of another by verbal means, language at its heart asserts *that something is the case*. Laughter registers that something is the case by registering its difference from what ought to be the case. This registration, however, is incompletely expressed: 'that' is not completely freed from inarticulate surprise – from the yelp, the dilated pupil, the raised paw.[5] In speech, what is the case, or what might be the case, or what is not the case (and hence by inference what is the case) is picked out, lifted up and held up to view.

Speech is a compendious, elaborate, folded, oceanic expression of what is there. What really or actually is there is presented as the fulfilment of a general possibility. Of course, in saying what is there, we say something about ourselves: we say what we are: we exhibit ourselves, we put into circulation an image of ourselves. When I tell you a fact, I tell you how well-informed I am, or how helpful I want to be, or how I feel about you. Every speech act expresses or invokes both a possibility in the world and something about myself. The speaking, talking, whispering, shouting, arguing, persuading, asserting, lecturing, soothing, flattering head points in at least three directions: outwards to the world, inwards towards itself, and outwards to an interlocutor.

Speakers use words. Words belong to all and everyone. Our mouths shape sounds thousands of years older than they are; our lips pat plosives that married their meanings on lips that have long since liquefied. Speaking, we exhale a mixture of air and history, breath and memory, beyond our consciousness. We speak with tongues not our own. And yet we make the language we have on loan our own possession, the most immediate and intimate expression of ourselves. Within a native tongue, a dialect, a culture, we carve out our *idiolect*, the mode of speaking that is unique to us. The choice of words, the structure of our sentences, the games we play with accents and sounds, and so on – all mark the common language as our private property. Beyond the distinctive pattern of kindness or cruelty, helpfulness or obstructiveness, ignorance and knowledge, gloom and joy, there is a special inflection of the semantic breeze, an idiosyncratic mix of imitations and echoes, that marks it out as ours.

And those words are realized in a voice that is unique to ourselves: 'the voice lies at the heart of what it is to be human'.[6] The tone and timbre, the music and dissonance, of the voice seem to give a hint of what it is like 'in there'. The 'grain of the voice' marks an intersection between the body, with its unique trajectory through the world, and the community of minds to which human beings also belong. The girl with evening tones in her deep voice, the penetrating mosquito whine of the resentful cockney, the Archbishop in whose throat syrup trickles over very smooth, very round pebbles: these voices broadcast the ambience of the country of the self from which they hail.

Boundless and fathomless, then, is the ocean of speech, the great sea of 'may be' constructed out of sculptured air, the ballooning space of possibility inflated by our ever-active mouths, where storms of abstraction bring joy and sorrow, war and peace, enlightenment and superstition; where a multitude of speakers stitches universes. By means of exhaled air, we can make prejudicial comments about Latvia when we are in Manchester, seek approval for our views on the Treasury model of the economy or a new shirt, curse the infidelity of our partners or sigh over rumours of war, rejoice in the beauty of a face or a distant planet. Using speech our shuttle heads weave a huge fabric; a world not of things but of facts.

Through the interactions of speech, the collective intelligence of a generation of heads can add up and each generation can be the beneficiary of the consciousness – the experience, the suffering, the discoveries, the wisdom – of previous generations. When, just a few thousand years ago, the collective heads of human-ity discovered writing, then the space of possibility expanded yet further beyond the human body, and the transmission of accumulated consciousness, sediment of the billions of streams of consciousness, from one generation to the next, became a gath-ering avalanche. In a very short time, writing has come so to dominate our lives that we might almost forget the connection between discourse and headwinds.

Linguistics experts tease out the miracle of speech in the phonology, the phonetics, the phonemics, the morphophonemics, the syntactics, the semantics, the semantic syntax, the pragmatics, the stylistics, the sociolinguistics, the physics, the neurology, the sociology, the anthropology, that together attempt to capture speech itself in more speech, in meta-breezes that try to utter what lies beneath utterances and makes it possible for them to have such precise meanings. And yet much of what others say bores us. Speech descends to wittering; and conversation is merely 'a vocal competition in which the one who is catching his breath is called the listener'.[7]

Our million-stranded vocalizations are time-soiled; the billion-threaded conversation the proliferating multitudes have with themselves is suffused with fatigue. We weary of speech littered with clichés, of the expected responses, the conventional utter-ances, the formal, the false, the approximate, remote from the dew of joyful realization. For talk never ends. When we are not talking to each other, we are muttering to ourselves. Our heads resound with expressed and silent thoughts. We live our final illnesses in words and descend to the grave murmuring to our many frag-menting selves and to those others from whom our seemingly endless conversations are to be taken for ever.

Ultimately we are no safer than speechless animals: the huge, many-layered bubble of 'that' in which we live, will pop. For the

present, we can keep it aloft; and so much knowledge and ignorance, so much sorrow and joy, is borne on the air our heads trap
for purposes quite unknown to the organic processes that led up to
the creation of our bodies. Our lungs would be nonplussed if they
knew what was happening to the stale air they were expelling.

Second Explicitly Philosophical Digression:
Enjoying and Suffering My Head

Nobody who has a headache or toothache, or who is feeling the warmth of long-awaited sunlight or drinking thirst-slaking iced water, can doubt that he is to some degree his head; and that he has to *be* it. Such experiences forcibly remind us of the non-transferability of our experiences. This aching tooth is something that we are obliged to endure and therefore in some sense *be*.

Well, not quite. As we have already noted, we are never precisely identical with any part of our body we may attend to. Nevertheless, the compulsory attention of pain narrows our distance from the painful part. It engulfs the attender, leaving little space for something that is both outside the pain and still part of the inside that is me. And this is particularly true if the painful part is in the head, as opposed to the foot. The aching tooth is up close and personal in the way that the toe, aching or otherwise, is not.

Suffering seems to be halfway between being and having; we say, 'I have toothache' but after a while we might just as well say 'I am toothache' or 'Toothache has me'. We move from 'I have this pain' to 'I am in pain' to 'I am pain'. At first suffering seems to be an interruption coming between me and myself. I go to sleep and hope it will pass away and I will no longer have to live it out.

Sooner or later, if it persists, it will no longer be outside of me: it will be me-as-outside. Sigmund Freud described himself towards the end of his life, after years of suffering with cancer of the jaw, as 'a small island in an ocean of pain'.

Hunger and thirst, which are indistinctly localized in the body, occupy the midway position between being and having with even more equipoise. It is interesting therefore, that while the supposedly stoic English say 'I *am* hungry/thirsty', the French say 'I *have* hunger/thirst'. The hungrier one is, the more the English seem to have got it right: as Henry Miller once said, a starving man is one large unsatisfied stomach. Nationals from both countries, however, agree on their relationship to tiredness. Perhaps because it is intrinsically diffuse and is not localized to a particular part of the body, we *are* tired, rather than *have* tiredness, on both sides of the Channel.

This aching head is me, and up close and personal, and yet the ache is impersonal. This is evident not only when we think of its origins, as when, for example, I am able to attribute it to a general process in my general body, such as the decay of a tooth; it is also impersonal in its manifest content. It is anybody's, or anyone's; and yet it fills every crack and corner of my individual being, spraying Paraquat over the personal, chosen, cultivated meanings that flower in the soil of my self. Because it has no one's name on it and yet I have to live it out, because it is not me and yet I cannot get outside of it, it is not merely an absence of meaning: it is positive anti-meaning.

Physical pain – which is at once intimate, alien and inescapable – is also an eruption of relatively unmediated sentience into the world, into the life, of the knowing self. That is why it seems at once closer to what we are than anything else and yet further from what we have become: the outbreak of anti-meaning (pre-meaning, un-meaning) is an alien landing but landing from within.

Hence, perhaps, the deep familiarity of pain. When I bang my head, the sensation connects me with the animal from which I have awoken, with the infant I once was.

Suffering, then, highlights our ambivalent relationship with the body that we-not-quite-are: at any moment, it could force an unchosen agenda of concern upon us, make us something we are not. And this is true even of the part of the body that is most our own, for example our mouths. That which is most intimately and most consistently your embodied self turns out to have properties that have nothing to do with you: daggers of otherness lie sheathed in our teeth. Even when we sleep alone, we sleep with a potential enemy.

Before we get too gloomy, let us remember cephalic delights: the pleasure of a fresh breeze on the cheeks when walking out on a summer morning; of a mouth filling itself with à la carte or the tongue of a loved one; of the head cradled in another's arms. Unfortunately, compared with suffering, all of these things, because welcome and indeed wooed and solicited, seem frail, tenuous. We fear our carnal luck that will not hold out. We are right.

chapter six
Communicating Without Air

They that have power to hurt, and will do none,
That do not do the thing they most do show,
Who, moving others, are themselves as stone,
Unmoved, cold, and to temptation slow...
They are the lords and owners of their faces,

William Shakespeare *Sonnet 94*

Air emitted from the orifices of the head is the richest, most elaborated, most powerful medium of communication between human beings. It has changed the earth. Nevertheless, the head may also communicate airlessly. For it is visible as well as audible and so can send out signals of immense significance without a word being uttered – albeit those silent signals often depend on a context established by words. The head may work as a whole, as in affirmative nodding or negative shaking; more often it utilizes part of itself in the service of communicating moods, attitudes, knowledge; informing, warning, manipulating, greeting other heads. Not for nothing did the German philosopher G. C. Lichtenberg describe the face as the most interesting surface on earth.[1] And we reward that interest with an acuity of attention of quite a different order from that which we pay any other surface. Our ability to identify and make sense of faces – to know whose face it is and what it is intimating – is extraordinary.

Infants learn early to pay a privileged attention to faces, amongst the mad kaleidoscope of sense data into which they are pitched. Newborns attend more to face-like stimuli than to anything else, though they do not discriminate between faces.[2] However, within a couple of months they are able to recognize individual faces: the presence or absence of mummy's face and its tone are of the utmost importance; it is the sun of the world. Eventually, we become expert face-readers, identifying one beloved or hated face from an archive of millions, reading its mental state, determining what it is attending to, and taking less than a second to see whether it is or is not beautiful or plain, attractive or unattractive, kind or cruel, amused or angry, ill- or well-intentioned, honest or lying.[3] We can even read the weather in the colour of the cheeks. And in almost as short a time, we can attach to the face a vast amount of biographical information about its owner; and an impressive dossier about the intersection of our own biographies with that of the owner, from the macroscopic facts that they are our parents, siblings, colleagues, to the microscopic facts that they were a bit off-hand when we last saw them.

'Attractive or unattractive': it is almost impossible to capture what it is that makes a face beautiful. What makes a melody soul-freezingly lovely so that we feel sunlit from within and sense that we live in a world radiant with possibility and that our lives are adventures? Such answers as are on offer – 'descending fourths' etc – answer nothing. Likewise, with the delicious texture and harmony and mystery of a beautiful face. A list of assets – large eyes, full mouth, smooth skin, a dark melancholy etc – gets nowhere near the sense, conveyed by such a face, that it promises or experiences another mode of happiness, a different mode of consciousness; that it hails from, or is the visible surface of, an invisible world remote from the visible world of ordinary days, ordinarily lit. Facial beauty hesitates between sight and sense, surface and significance – which is why novelists very quickly drift from anatomy to hermeneutics, from the physical to the personal in describing their characters. They use terms such as 'animated' to describe beautiful eyes, or rosy cheeks, and in so doing collude with the world that muddles its desires with moral judgement. A spotty face may be attributed a spotty soul and a scarred face a scarred soul; and this may be self-fulfilling. It takes a special grace, after the blow that created the scar, to resist the further blows of the curious, averted, undesiring, cruel gaze of others.

Identifying a face in a crowd – which involves separating small facial surfaces from the larger canvas of sensibilia that the world presents, and picking out the one face from others – is a challenge. There are dramatic impairments of facial recognition – which the neurologists call 'prosopagnosia' – which may occur after a stroke or brain injury affecting part of the brain that coordinates facial perception and memory.[4] In some cases, people are not able to recognize even the faces of those who are closest to them; needless to say, this is deeply disturbing and socially crippling. There is another condition, the so-called 'Capgras syndrome', in which there is a loss of emotional responses to familiar faces. This leads sufferers to believe that someone close to them is an impostor, or a double. And then there are ordinary failures, when we pause for a moment over the face of a stranger and realize that he or she is our spouse. That moment of hesitation makes us aware of the dissociation between an individual's identity and that by which they are identified.

Identity and identification. Two terms sounding so similar and yet so far apart in meaning. Their profound difference is particularly evident in the case of our selves. That which I am, that which I feel I am, and that by which I am identified, barely touch. This is not just because I look *out of* my face and you look *at* my face, not simply because my face is transparent to me and opaque to you: no, it is because of the dissociation between the standing structures of the face, the current account of the moment, and the self extended in time. Our face and our soul's CV are only contingently connected with one another. So-and-so's sweetness of temperament does not modify the plainness of her face that leads others to withhold desire, admiration or even interest. So-and-so's malignant destructiveness does not make the letter of recommendation that is his handsome features the slightest bit less calligraphic, any more dog-eared, or less persuasive. That is why, as Shakespeare noted, 'There is no art to read the mind's construction from the face'. All faces are to some degree poker.

Even so, we are more or less identified by our faces and to some extent identified *with* them. If, as Wittgenstein said, 'the human body is the best portrait of the human soul', then the face is the essence of that portrait – as portrait painters acknowledge. Many portraits are of the face alone; a portrait that consisted only of a pair of legs, a buttock or a spleen would be a joke. You might point to my face and say 'That's Raymond' but if you pointed to my leg you would say 'That's Raymond's leg'. In Oxford there is a stretch of the Cherwell – Parson's Pleasure – set aside for naked male bathing. Once, in less relaxed times, a group of ladies accidentally went off course and punted past. Embarrassed dons seized their towels and, with one exception, clutched them to their private parts. The exception was Maurice Bowra who placed his towel on his head, observing that 'In Oxford I am usually known by my face'.

The head's mode of communication without air is almost as large a topic as speech, so I have selected three key examples: nodding, winking and smiling.

A Nod and a Wink
We are all familiar with Ministers sitting behind the Prime Minister at the despatch box at PM's Question Time, frantically signalling their agreement with what he is saying. Of all the nodding

rear-window doggies, Dr John Reid, former Home Secretary, was the most active and will go down in history as one of the greatest dittoheads of the twenty-first century. As doctors, we are instructed to nod at regular intervals, to show not only that we are listening to, but that we are in some non-specified sense agreeing with, what the patient is saying, at least to the point of understanding and sympathizing with it.

Anaerobic head signals can convey quite abstruse messages if things are set up right. Which should perhaps astonish us more than it does. After all, nothing could be simpler than a controlled loll-and-recovery. With its assistance, we can sign our agreement to a comment about economic trends, support admission of a new member to a club, acquiesce to a request of the most or least significance. Nodding can be stitched into the most abstract or concrete, history-laden, personal or impersonal, formal or informal, of assertions, petitions and claims. Many things can be passed on the nod, or nodded through, but only as a result of a huge amount of ground-work. The slightest nod reaches deep into a hinterland of a multitude of concepts, frames of reference, and worlds.

No wonder nodding has so many inflections and dialects and tones and volumes. There is the rapid nod of a man impatiently agreeing with something we are painstakingly spelling out, or acceding to a timid request that has been granted. This is the 'Yes, get on with it' nod, which may be further abbreviated as the 'curt' nod. There is the slow, long sweep nod, which encourages its recipient to accept or agree to something that is being put to him. I may highlight my nod by staring at you with an unnaturally wide-eyed gaze, which makes clear that this nod is not a tremor or tic but a deliberate communication. There is the minute nod, which cloaks itself not only in silence but also in invisibility. I place my positional advantage at your service so that you may cheat at cards. And that is just for starters.

Nods are often partnered with winks. The Martian's guide to winking – who winks to whom, under what circumstances, and why, and what is winked – would fill many volumes. The versatility of the wink – a transient, monocular occlusion of vision – or more precisely of the winker, is rarely fully appreciated. Winking may send lighthouse flashes across generations and across the sexes, communicating not just some information I wish to impart but also my attitude towards you.

We wink at 'youngsters' to signify non-specific parental or quasi-parental benignity. We may show solidarity with them, when they are oppressed by a po-faced, bossy, demanding world. Winking is a momentary challenge to the inner parent by the inner uncle. Winking says, 'Don't worry, I am on your side, even though the world is not.' Winking may also say, 'I am not serious, this is not serious' or 'Don't worry, don't take it to heart'.

Winking across the sexes is altogether more hazardous. Winking at grown women may invite a verbal or physical slap. Its object can feel patronized. More than that, the wink attempts to sexualize by establishing an asymmetrical situation where the female is more the object of the male's consciousness than the male is the object of the female's consciousness. Winking also has a foot in the realm of that great territory of the risible, where sexual desire, which wishes to put a hand up a skirt, has to advance through a fully dressed world where intercourse is largely verbal, formal, practical and asexual.

When the female is the winker, we are in yet more dangerous territory. The comedy wink cries 'Hello, sailor' and takes the sexual initiative. It says. 'Come, I know what you're after. Why not take it?' It also mocks male winking and regains the initiative: the winking woman is knowing, rather than merely an object of knowledge, or of desire-fuelled quasi-knowledge; she is a subject rather than a mere object. The winking woman winks at the history of winking and fights back against the helplessness of women caught in the male gaze.

Winks, for all that they subvert seriousness, are clearly a matter of considerable seriousness. No wonder we seemed tuned to detect them, despite their minute size. They range from a mere blink to a full eye squeeze to a wink and a nod of the head. The accompanying nod is interesting: the oblique, rotatory movement of the head that goes with the hemi-facial spasm is halfway between an affirmative nod and a negative shake. That's winking for you: ambivalent, questionable, uneasy even in the means it uses to raise its font to increase visibility.

Calculated winks show how far we go in transforming spontaneous events into deliberate actions, reflexes into calculated signs. No wonder, then, that many philosophers have used the distance between winking and blinking as a marker of the difference between determinism and free will, between humans as organisms and humans as more or less conscious agents. To take the blink that protects the eyes and divide it into half; to enhance it in the way we have noted; and then to use it to transmit a message in a context that has been set up to disambiguate it, is to highlight one's distance from the organic world, from the *donnée* that is one's biological body. In deciding to wink, we mobilize our past experience, a great slice of the present steeped in cultural history, a gradually evolved and painstakingly acquired symbolic understanding, and, by means of this great lever, we operate on the massive given of the body to subordinate it to our aims, to our goal of forging a distinctive kind of presence in the world.

Not all winkers are carnal, of course. The winker on the car is an egregious indication of my intended direction of travel. It is pointing pared down to a minimalist glance, exploiting the fact that the intermittent is more attractive of attention than the sustained.

Smile

> You move the 57 muscles it takes to smile.
>
> Michael Hofmann, 'First Night'

The effectiveness of winking is a reminder of how minutely we decipher faces – however small the print, however Gothic the script, however unsatisfactory the light that falls between your face that writes and my eyes that read. According to Paul Ekman, a social psychologist who has devoted a lifetime to the study of facial expressions, there are seven basic emotions expressed on the face in the same way in every culture: sadness, anger, surprise, fear, enjoyment, disgust and contempt.[5] These are innate, not learned; which explains why congenitally blind people show the same facial expressions attached to each of these seven emotions as sighted people. The expressions may be very brief: so-called 'micro-expressions' – important evidence to a doctor of an emotion that is being suppressed or concealed which may last as little as a fifth of a second.

How important all those expressions are. They tell us about the current account of your soul, what is happening between you and me, what kind of person you are, what kind of person I am. They exploit the wonderful mobility of the facial muscles, which arises from the anatomical fact that, unlike other muscles, they are directly connected with one another and not tethered to bone: they are rags fluttering in the breezes of inner and outer meaning. According to Ekman, there are forty-three muscles involved in facial expression. Like rural villages, the smaller they are, the more exotic their names: consider the minute Levator labii superioris alaeque nasi (the elevator of the upper lip and the wing of the nose) which is responsible for dilating the nostril and elevating the upper lip. Ekman has developed the Facial Action Coding System which identifies 10,000 visible facial configurations of which 3,000 are meaningful.[6]

Our exquisite sensitivity to others' facial expressions is especially striking in our ability to detect, categorize and interpret the smiles we exchange with our fellows. This is as well for, as Angus Trumble reminds us, the smile is 'the most immediately expressive muscular contraction of which our bodies are capable'. It is a primordial communication: two-month-old babies smile at their mothers and seem to mean it. The smile in portraits offers a 'convenient shorthand for far-reaching assessments of character, behaviour and temperament'.[7]

Trumble classifies his smiles: decorous, lewd, mirthful, deceitful and wise. But this is simply the beginning of a classification of facial events that have infinite subtlety and shade into so many modes of not-quite-smile – the grin, the smirk, the grimace – and which are in many instances unclassifiable. We observe of a smile whether it is warm, or cold (an especially shocking index of the cruel and calculating soul within). Consciously or unconsciously, we register whether it is an affair mainly of the mouth or engages the eyes and conclude accordingly as to its sincerity and warmth. We see that it is spontaneous or forced, confident or shy, disinhibited or demure, sweet or sour, guileless or enigmatic.

No wonder so many spend so much time looking at their faces in the mirror – 'preparing a face to meet the faces that they meet'.[8] We may judge our own smile as ugly, or even false. As a child, I longed to transform my giggly, silly, immature grin into a weary wisdom and loved, in the solitude of my bedroom, alternating between running the culpable one and the laudable other. The ideal was the Chekhovian smile that looked upon the world with sorrow, sympathy and understanding: the smile of the foible-free upon a human world of foibles.

The relationship between smiling and laughing is neither clear nor happy. Their connection is underlined in French, where *sourire* would appear to be a suppressed mode of *rire* – a kind of 'under-laughing'. Smiling and laughing may be seen as alternative responses to the same stimuli, the choice of one over the other being a marker of sophistication or even social standing. 'The vulgar often laugh but never smile,' the Earl of Chesterfield noted on 17 February 1754, 'whereas well-bred people often smile, but seldom laugh'.[9] Whether this is epidemiologically sound is not certain but one can imagine why he made the claim. Smiling does not expose one's bad teeth, emit noxious gases, make loud noises that invade the acoustic space of those around one, result in urinary incontinence (no one pissed themselves smiling) or leave the face discomposed and vilified by redness, tears and the other stigmata of a prolonged laughing jag. More profoundly, the smile registers a more private and subtle amusement that does not seek external confirmation; it is about the self-sufficient 'I' rather than the mutually supportive 'we'. We laugh at that which others find funny, we smile at things that amuse ourselves. The vulgar, less possessed of independent substance than the well-bred (whose clothes, possessions, lands and rights were internalized as existential substance), would seek to recruit others to the cause of their amusement to validate it.

Smiling reaches beyond mirth: it is not simply about registering the gap between how things are and how one would like them to be. Cruel, joyful, loving, triumphant, resigned smiles cannot be gathered under the normative activity of a human consciousness trying to 'make the world his thing'.[10] We may smile with pleasure, friendliness, with greeting, and agreement. We cannot therefore see smiles as merely less intense, or anaerobic, laughter. The modulation between modes of smiling is unmatched in laughter: the passage from a grin to a sneer, from a smirk to a concessionary smile, from tiredness to a world-weary smile, cannot be mapped in any of the journeys between sniggering and tittering, between tee-heeing and braying, between suppressed pantings – stuffed into the body as into an over-filled suitcase, so that the whole trunk shakes – and full-blown guffawing.

Smile-watchers seek out subtleties at the opposite pole from the full-face, ear-to-ear beam. The point where the professional seductress's enigmatic smile wanes to weariness. The smile on the cadaverous face of the dried-up pedant like (to borrow from Dickens) 'sunlight on the brass handle of a winter coffin'. The pillar box slit of one whose worst expectations have been fulfilled. The moment when smiles appear through tears and weepers surprise themselves with changing feelings they read from their own 'watery smile'. (Such a phrase makes one smile at the genius of ordinary language.) Twinkling smiles call on the play of the eyes that have more to do with light than the lips or the teeth. Smiles may be deemed to be wry, ironical, mocking, self-mocking, malicious, triumphant, gloating, satisfied, self-satisfied, sour, sickly, sardonic, restrained, patient, forbearing, indulgent and generous. Their intensity may be translated into temperature (warm, cold) or accorded more abstract dimensions, particularly when it is their sincerity that is at issue: hence, the wan, the forced, the set, the fixed, and the frozen. There is the empty smile to which no significant feeling corresponds:

> Eternal smiles his Emptiness betrays
> As shallow streams run dimpling all the way.[11]

Smiles may be spontaneous, especially appreciated when they come from a structurally scowling face, like sunlight through a break in the clouds. Or they may be calculated or calculatedly spontaneous. Or assumed in order to unnerve, to make the other feel uncomfortable. To be stared at is bad enough (as we shall discuss in chapter 8); to be subject to an unexplained smile is much worse. Is there egg on my chin, have I said something stupid, have I betrayed something inadvertently, do they know something about me that I do not?

Royal smiles are especially valued. One royal, the late Queen Mother, almost became her smile – like the Cheshire Cat – and was driven over great tracts of the surface of the earth so that her daughter's subjects could see and cheer it. The smile, the epicentre of her presence, was always described as 'radiant' – no Geiger counter could withstand it – notwithstanding the brown teeth that it unmasked.

At the opposite pole to these very public smiles, there are those in which we smile to ourselves, thinking of the past, of something to look forward to, with satisfaction at some achievement or a distasteful task done, at the beauty of a logical proof or a melody, at a joke we have heard or invented, or a machine that works, the 'aha' smile, the smile of satisfaction, when we have suddenly understood something, or seen how to do it, or seen how it was done. The mad, traditionally, smile, as they laugh, to themselves more than they smile to others. This is less a sunlit communication to the outside world than moonlight beamed from one bit of their world to another.

In the littoral zone, there are fleeting smiles, shadows of smiles, most elusive of all, the ghosts of smiles. Think of that: the ghost of a smile – an adjective, an adverb, becomes an object and this object is haunted by the past, the idea, of itself. And there are suppressed smiles; and the 'tongue-in-cheek' smile, described thus on account of the move taken to control one's lips, to conceal the Mickey that is being taken.

The most life-enhancing, destructive, hope-dispensing and despair-making metamorphoses of the smile surround love and sex. Shy, embarrassed, warm, come to bed, come hither – such are its modes. In the warmth of the sexy, inviting, welcoming smile there is concentrated the warmth of the body, addressed to you, and you only. Anticipated pleasure that only you can confer glows in that smile; and the satisfaction of pleasure received where delight and gratitude meet in the smile of love. The joy is great; and, when the smile is withdrawn, the catastrophe proportionate.

Smiles are intrinsically volatile and a source of endless surprise. Think of the schoolboyish grin suddenly flashing out in the midst of the most serious or tense situations. Mr Blair, our late Prime Minister, was a past master of this particular facial *volte-face*. His credentials of seriousness are beyond challenge. The Prime Minister is the most adult person in the nation. And there he is, at Prime Minister's Question Time (when the country's most adult questions receive the country's most adult answers), thinking up a riposte to an opponent. An idea comes. He is tickled. There flashes out – beneath and beyond the triumphant put-down – the famous schoolboy grin. This is the grin of the pupil on the verge of trouble, of laughing at something that perhaps he shouldn't, or not at least at that moment, or in danger of laughing too much. He is one of us. He is human after all.

It is hardly surprising that damaged smiles are so socially damaging and, given that they are so minutely inflected, how vulnerable they are to damage and their signals reduced to permanent toad-speak. The half-smile of Bell's palsy, in which the nerves supplying the muscles of expression on one side of the face are put out of action, looks like a marriage between a shocked face and an independent demi-rictus – a smile divided anatomically rather than diluted by lack of warmth. It is then that we find out who loves us for what we are, who are our real friends.

More morale-undermining still is Parkinson's disease in which the initiation of movement, and control of change of direction, is impaired. The sufferer has the most grinding form of poverty – poverty of facial expression. His set, unsmiling face suggests sullenness or emptiness, granite unmovedness within, a state reinforced from without by others, who, not liking their own smiles to go unrewarded, withhold them. What is more, the patient is deprived of the retroactive effect of one's own smiles that carry good cheer back to their source. The infrequency of smiles brings about depression.[12] No wonder November seeps into the soul and the leaves of the world are crushed to coal black misery. Smile and the world smiles with you; weep and you weep alone.

chapter seven

Notes on the Red-Cheeked Animal: The Geology of a Blush

Man is the only animal that blushes – or needs to.

Mark Twain, *Following the Equator*

For Epimetheus I was called by my progenitors,
he who muses on things past, and traces back,
in the laborious play of thoughts, the quick deed
to the dim realm of form-combining possibilities.

J. W. von Goethe, *Pandora*

The human face is the most sign-packed surface in the universe. Scowls, grimaces, curled lips, raised eyebrows, furrowed brows, blown out or sucked-in cheeks; quizzical, arch, loving, suspicious, contemptuous, thunderous, forlorn, frightened, weary looks; disgusted winces, set-mouthed rage, diffuse haggardness – these are just a small sample of the sheer range of the expressions that shimmer in endless procession across our faces. But there is one singular and wonderfully expressive facial sign that deserves our particular attention: the blush.

Amongst other things, blushing is a reminder that much of the meaning that emanates from our own faces, and that we read into other's faces, is not actively *meant*. First, the way our facial expressions look to others is, notwithstanding all our calculating and uncalculating, kind and selfish, self-consciousness, almost as little under our control as the structure of our face and the extent to which others do or do not find us attractive. Secondly, not all expressive facial events are voluntary. While nodding, winking and smiling are for the most part voluntary (they would not otherwise serve the function they have), there are other events – tics and so on – that are involuntary and paramount among these is blushing.

The neural basis of blushing has proved difficult to pin down because of the peculiar difficulty of investigating blushes. The use of volunteers of a certain age and a certain sex does not always help. This is illustrated by the experiences of a researcher with a series of young females. Attempts to prompt blushes under experimental conditions by exposure to suggestive material were unsuccessful. When, at the end of the abandoned session, the experimenter thanked the subjects for their help, they said that was quite all right, apologized for their uncooperative cheeks and blushed scarlet.

More recent research has demonstrated that blushing is due to a combination of factors. The most important are the unique sensitivity of the facial vein (that supplies the small vessels of the face), which dilates rather than contracts in response to adrenaline; and the anatomy of the blood vessels of the cheek, which are denser, wider in diameter, nearer to the surface and less obscured by tissue fluid.[1]

So much for the physiology. The psychology is even more complicated and elusive. Blushing is associated with undesired social attention and heightened self-consciousness. We blush with embarrassment, with shyness, with uncertainty, with a sense of exposure, of undress. It first becomes common in children of kindergarten age, when the head's awakening out of, and to itself, has become elaborated into a 'social self'; and it peaks in adolescence, when social anxiety and self-awareness also peak.[2] Its function is not clear, particularly as it draws even more attention to the person who is already excessively aware of being attended to: blushing broadcasts our sense of being undressed. It has been suggested that it is a 'non-verbal means of saving face', of mitigating the negative reactions of others – a pre-emptive physiological *mea culpa* – and studies have shown that people react less harshly to mistakes when the perpetrators blush.[3]

At any rate, it is a godsend to the sexually predatory. While making a pretty face blush may or may not enhance its prettiness, it sexualizes it by bringing its warm physicality to the surface. The low-grade sadism of making a shy girl blush with some smutty remark is a step in the direction of exploring the spaces under her skirt. Irrespective of whether, according to Freudians, the reddened cheeks are a proxy for engorged genitalia, the hectic flush makes the vasomotor instability, the body's self-experience, of a nervous girl more graspably present. Her burning cheeks upgrade her would-be seducer's gaze to a burning glass; the living rouge, seasoned with an equally helpless titter of embarrassment, is a satisfying mark of her vulnerability, even of her wanting, notwithstanding her protests to the contrary, what her suitor wants.

In order to avoid such distractions, I want to focus on a decidedly non-sexual blush in a person of the male sex, because blushing is a kind of glass-bottomed boat enabling us to look at the depths upon which our ordinary moments float. It provides a plausible pretext for a direct exploration of one of the central themes of this book: the great gaps that have opened up between humans and animals. We noted how the secretions that come out of anatomically separate, physiologically solitary human heads become incorporated into a nexus of signs, an explicit common culture. Unlike animals, we belong to, have our moment-to-moment being in, a community of minds. Entry into this community is via a self-consciousness that begins with our assuming our bodies as our own, as ourselves. This self-consciousness becomes intricately folded and massively elaborated as in the first few years of life we grow into, assimilate, and become assimilated by, the pooled consciousness of humanity. Almost from the word go, we are not to be understood, as animals may be understood, as stand-alone organisms; even less are we to be understood as stand-alone brains.

The story of a single blush is one way of getting the measure of the journey taken by human consciousness once hominids awoke from their organic state. The story also, and incidentally, breaks down the artificial boundaries I have set up between airborne and airless communication, as it weaves between air present, ancient air, air breathable and air turned to written words as enduring as granite. At any rate, it provides an opportunity to look again, and in more depth, at some aspects of speech – that mode of explicitness which, though a relative newcomer, has, more than anything else, transported our heads to places unknown to any other organ in the animal kingdom.

The Occasion

My specimen blush blossomed on the cheeks of forty-four-year-old Charge Nurse Ryan, my colleague on the Elderly Care Ward. (I have changed his name to spare his – well, you know what.) It was prompted by a comment he made at a meeting we had convened to determine the best approach to decanting patients from one ward to another: from H1 ward, which was about to be painted, to H2 ward which was to house them. In front of the assembled team, Charge Nurse Ryan referred to the second ward as 'haitch 2' ward. The superfluous meta-aitch ignited a blush that spread to the roots of his hair. I could see that he was squirming, sweating, puddled in the shame of self-betrayal. Though nobody apparently noticed, I fear that he inferred that everyone did.

I wanted to help him and I did so by briskly moving the meeting on to more detailed logistics. I did not, for example, try to diminish the small accident of his embarrassment by locating it on a larger canvas. I did not invoke the great stories of evolution, of prehistory and of history, in the hope that, like the focal crimson of dawn in the edgeless hours of daylight, his pinkness would fade in a widening astonishment. This, however, is precisely what I would like to do now: to assume the role of Goethe's Epimetheus – one who 'traces the quick instant to the dim realm of form-combining possibilities' – by reminding myself of the remote aquifers of consciousness from which contemporary blushes may sometimes draw their pinkness.

Speech

In the beginning there was speech. When that beginning was, no one knows. The aphasia of the material world, and of the experiences of the sentient beasts in it, ended perhaps forty thousand, perhaps a hundred thousand, perhaps several hundred thousand, years ago. Nobody knows how it came about that speech broke into the wordless universe, though there have been many theories. None of these comes near to crossing the gap between the calls of animals and the discourse of humans. As we have noted in chapter 5, few of the theories even see the size of that gap. Let us try at least to measure it.

Many things have meaning – clouds, smells, spots on the skin – but only a few of those meanings are meant. Human speech is, to a unique degree, 'meant' meaning. As such, it proposes mainly abstract *possibilities*. It achieves this astonishing sculpturing of 'may-being' by postulating referents that are captured by a constellation of general senses. Referents are not obliged to be real. Even less are they required to be present. Speech-conjured absences come in many commensurable and incommensurate modes: 'no longer', 'not yet', 'never'; out of sight, out of ken; subjunctive and optative; probable, possible and impossible, etc.

The point is that none of this can be hissed, barked, squealed or hooted. Even less can it be rustled, rumbled or thundered. The awakening of speech was a cognitive explosion in the inarticulate universe. The first person to say 'goo' to a goose opened up new eyes in nature so that it began to see itself. The foundations for that self-consciousness and awareness of social attention that makes humans uniquely blush-prone.

Reported Speech

Some time after they started speaking, talking hominids learnt the trick of reporting what others have said: X says that Y said such-and-such; Z quotes herself.

Reported speech, which is not merely an echo or imitation but genuine present talk *about* past talk, purposeful replication of what has been uttered, marks a further immeasurable distance from animal calls. It reflects something we noted in chapter 5: how human language, unique among animal communication systems, is riddled with meta-language. Meta-language expresses, and creates, ever higher orders of consciousness. The attribution of speech, either to another person or, indeed, to oneself, is a striking manifestation of the extraordinary awareness humans have of each other. When you say, or remember, or even erroneously claim, that I said such-and-such, you reveal your complex idea of me as a very special kind of agent, one of whose activities is to emit words that I *mean* in order to create or convey (quite abstract) meaning.

We humans, in short, know that other humans are the sources of meanings that are *meant to be meant*. This confers great powers upon us, so that we are able to bring about events of quite a differ-ent nature from those that are mediated through the boundless unselfconscious energies of the material and biological world. Such powers were unknown throughout the first 99.998 per cent of the lifespan of the universe. And now they are ubiquitous; so much so, that one could be forgiven for the anthropocentric delusion that the greater part of the universe is the shared world we have woven out of meant meanings.

This explains why our nakedness is at once more complicated, more folded and yet more naked, than that of pebbles or, indeed, of any other living or non-living object in the universe. We have an enormously extended surface of exposure. Charge Nurse Ryan's red cheeks – hot, unbearably self-present – have a venerable ancestry. Though he felt alone at that moment, he was not alone in history. The occasion might have been unique but not his dis-comfort: every individual who speaks and meta-speaks, and therefore has a folded consciousness, runs the risk of red cheeks. While the rubefacient may vary from person to person and from time to time all humans are blushworthy.

Inscription

We are taking the long view and trying to trace the winding journey of awakening that matter took on its way to Charge Nurse Ryan's blush. The next step is writing or, more broadly, *inscription* of speech. Inscription liberates meant meaning from warm mouths and body-heated air, and widens the circle of interlocutors without limit. By means of the unfading written word, the kitten-prints of thought are set in concrete: ideas can reach across the world and echo down the ages.

Writing is a relative newcomer compared even with that parvenu human speech. A mere 9,000 years old, it has featured in only 0.00005 per cent of the history of the material universe. Knowledge, incidentally, of the relative longevities of the universe and of writing is itself demonstration, if demonstration were needed, of just how far inscription allows human consciousness to reach beyond itself! Of how the collectivization of human awareness through the written enables each mind to transcend its insect body; or, at least, to envisage a boundless world that does.[4]

Writing does not capture thought or experience directly. It is second order language: its signs are conventional signs of conventional signs (in contrast with gestures and pictures which, even if conventional, tend to be first-order conventions).[5] No wonder writing was such a long time coming. The wonder is that it came at all, especially since script is not direct *transcription*: we rarely speak as we write, notwithstanding that the primary task of writing is to symbolize the symbols of language by indirectly capturing the sounds of words as they are usually used in speech.

Alphabetization

Our story of the aetiology of Charge Nurse Ryan's blush is still far from complete. For a start, he would not have caused himself to blush in precisely the way that he did had the genius of the Semitic peoples not introduced the alphabetic principle and by this means further divorced inscription from the voice. Written words, unlike the spoken, would henceforth be made of *letters*.

Not surprisingly, alphabetization was a gradual process that passed through many stages. Picture-writing, in which the written word was a more or less conventional image of its referent, was replaced by word-based writing systems, where whole words were represented by single signs. The great breakthrough, enabling writing to make infinite use of finite means (to borrow the linguists' awestruck expression – and which applies to so much of what the human head does), was the invention of *sound-based* syllabic writing systems. In such systems, signs were used to represent a common sound rather than a common meaning. Finally, only 3,000–4,000 years before our index blush, syllables were broken into their constituent consonantal and vowel sounds. This was the birth of letters proper.

Such leaps – aided by transitional forms, such as acrophonics that represented a sound by pictures of things whose names began with that sound – are unimaginably brilliant. Every step was a work of genius; even the smallest step, as when those clever Greeks used the spare consonantal signs left over from Semitic inscription (because they did not correspond to any sounds in their own language) to stand for vowels, and thereby completed the alphabet, was in truth a huge leap over a high sill.

And all of this took place 1,000 years before God, according to some people, was minded to inscribe his own Word in the humanized flesh of his Only Begotten. Just in time. Had script arrived much later, the Crucifixion would have been a fading memory. Instead, as it has turned out, it has proved the justification for relocations of blood rather more radical and on a larger scale than blushes.

Spelling: Sounding the Letters

Written words could now no longer be thought of separately from the alphabet: they were identified with certain stipulated letters placed in a non-negotiable order. And so there came into being the convention of the correct *spelling*.

Humans took spelling lessons. Such lessons involved repeated 'spelling out'; more specifically, CN Ryan's nemesis: *spelling out loud*. Spelling out loud, enunciating the right letters in the right order, requires some vocal means of picking up letters. A mere echo of the sound that the letter contributes to words would not suffice as the necessary tweezers; for a given letter may contribute a multitude of sounds to different words. For example, 't' plays different tunes in 'tune' and 'the'. So we give formal names to the letters, quite distinct from their sounds. To add to the confusion, and another twist in the road leading to CN Ryan's red cheeks, these names themselves have to be sounded. What is more, the letters that stand for sounds are named by sounds that may in turn be composed of letters – so 'tee' is composed of 't' plus 'e' plus 'e'.

'H'

Our ascent from the depths is now almost complete. We are getting close to that daylight in which CN Ryan experienced his nakedness in the averted gaze of those others who, he felt, registered his blunder. His blush requires just a few hundred years of further preparation. Before we reach the open air in which his red cheeks glowed, however, we need to rehearse the fortunes of the mysterious letter 'h', aka 'aitch'.

For the phonologist, 'h' corresponds to, or signifies, a simple aspiration or breathing before a vowel, with just enough narrowing of the glottis to be audible. (Try it out.) I like to think of it as the letter closest to the voice, to the airways, to the lungs, to the living body of the person who says 'I am'; to the *pneuma* of the living spirit. Less sloppily, less romantically, it strikes me as a hybrid form, halfway between vowel and consonant. Unlike that all-rounder 'y', which flips between the two roles, 'h' fails properly to be either: it has neither the fibrous toughness of the consonant or the ballooning openness of the vowel.

This indeterminate state may explain why the name for 'h' (or 'aitch's' name) is so odd. It is remote from any sound the letter actually contributes. When we call 'o' 'oh', the name forms at least part of what its object sounds like, reproduces one of its roles. And the name for 'a', too, sounds one of the notes that 'a' sounds when, teamed up with the right consonants, it is doing real work. And 'pee' at least begins with the sound that 'p' makes. So how did 'h' get to be called 'aitch'?

'Aitch', it transpires, is the last stage in a thousand-year-long road which the name of the eighth letter of the alphabet passed through, taking in the late Latin 'aha' (which at least includes the sound in the name) and the Middle English 'ache'. This journey, by a wonderful paradox, involved 'h' running out of breath, losing the very spirant that it is itself. When 'aha' became 'aitch', its 'h' was buried in an unaspirated syllable.

Historical Sociolinguistics

So much for the puzzling name. But there is another story of 'h', that reminds us how often what speech does goes beyond (or may have very little to do with) its referential, or factual, content. To pick up on this, we need briefly to step back from the written word and return to the spoken one.

'H', I am sorry to say, is a snob. It prefers the mouths of the rich and titled, the comfortably off, the white-collared, to the poor and tiaraless, the hard up and the blue of collar; it favours those who work with abstractions above those who move matter, those who run offices to those who till fields. It is an important marker of social position and consequently speaks directly to the redness of the cheeks of those red-cheeked animals who nurse upward aspirations: those examples of *H. sapiens sapiens* who would dye their blue collars white, and wish their hands less horny from friction with the material world, less expressive of the work of the body than of the work of the mind.

As the *Oxford English Dictionary* puts it (disapprovingly, as if it were itself an entirely innocent party) 'the correct treatment of the letter H was established as a shibboleth of social position'. Though every word we utter is to a lesser or greater degree a shibboleth, enabling others to type or stereotype us, the absence of 'aitch' has pre-eminence.[6] Missing an 'aitch' off a word, omitting this elfin's expiration, this pico-pant, betrays your origin: people who do not drop their aitches can place you at once and know you are not one of them. The dropped 'h', a next-to-nothing still-born as an audible nothingness, is a legendary unmasker of the social climber, marking the moment when the hand-holds pull away from the wall and the abyss opens. If those others had for a moment imagined that you were one of them, they now know that your membership is invalid, because it is based upon external show, not breeding.

In traditional comedies of manners, this is often because your money is too new. Eventually money becomes breeding, for time launders loot, but you have to have enough of it. Lucre you haven't earned yourself is clean; and being brought up among the accoutrements of the well bred turns in a very few generations to *being* well bred.

The absence of 'aitch' is important enough to be tactlessly flagged by apostrophes in novelists' and others' 'realistic' accounts of vocal presence. Aitchlessness or aitchopenia may condemn the speaker to walk-on parts, cameos and character roles rather than leads. The aitchless are off-centre: rural and remote rather than metropolitan and near; in the street market and the field, not the chandeliered room; in those places where the effects of power are suffered rather than brokered; where humans are ordinarily conscious but consciousness is not at its highest.

The Blush

We come (at last) to Charge Nurse Ryan's rubefacient error.

In the course of the wonderfully exotic act of spelling out a word by uttering the names of its letters, he added an aitch at the beginning of 'aitch' where, of course, there should be none. His 'haitch' was no exercise in historical phonology, an attempt to restore to aitch the aspiration it had lost in the Middle Ages. No, his redundant 'h' betrayed the influence of a pervasive anxiety at the possibility of dropping an 'h' where there should be one. It was an over-compensation fuelled by an exquisite awareness of how loudly the dropping of a something so light as a pufflet of emitted air, an oral micro-zephyr, may sound. Deploying the laryngeal guttural spirant or rough aspirate where there should be unvoiced smoothness betrayed the aspiration to smoothness of one who might be deemed by some as rough. Cheek-scorching shame ensued.

His anxiety, though experienced individually, was transmitted from antecedents unhappy with their social being. His 'haitch' belonged to the dialect of the upwardly hopeful, doomed to be arrested in their ascent. The redundant 'h' caught the escapees, who wished to rise without trace, red-cheeked. It revealed not what they were but what they wanted to be; or uttered the distance between the former and the latter. His blush, in short, expressed a collective shame on behalf of his tribe. And a betrayal of the tribe. And so a self-betrayal.

Conclusion

We have encountered a succession of mysteries: speech, the reporting of speech, inscription, alphabetization, the dis-traction of words into separate letters, the naming of letters and the uttering of those names, and social self-awareness mediated through the sound of one's own voice. These are mysteries that lie deeper than any blusher could be expected to feel, and reveal how far humans have had to travel to make such discomfort possible.

chapter eight
The Watchtower

Ineluctable modality of the visible.

James Joyce, *Ulysses*

First Glance [1]

Sometime between 4 million and 8 million years ago, the weather in Africa took a turn for the worse. There was a drop in global temperature, resulting in cooler, drier conditions. The earth became less hospitable and there was a loss of tree cover. Our primate ancestors were expelled to the plains. Life in the relatively treeless plains favoured the upright position: there were fewer trees to obstruct upright walking or to provide support for brachiating from branch to branch. Hominids learned to walk on two legs.

Although other primates assume the upright position from time to time, only humans are overwhelmingly bipedal. What is more, humans alone are capable of *striding*. At each stride, the arm opposite to the forward leg swings to compensate for the twist of the trunk, enabling upright walking to be more stable. The upright position elevates the head to a bust on a long plinth. This is of particular advantage for the distance senses – such as vision and hearing – and increases their importance compared with proximate senses such as touch and smell. The elevated head, with its greatly enhanced visual capability, makes the body a kind of watchtower. Vision consequently accounts for about 90 per cent of the information that we acquire about the outside world through our senses. This has had innumerable consequences for our relationship to the natural, and more broadly the material, world around us, for our relationship to other human beings, and for our sense of ourselves.

I will examine each of these in turn. But first, it is appropriate to say a little bit about the eye – which is so intimately bound up with our subjectivity – as an object.

To say that the eye is 'a complex organ' is to understate things: a large number of minute parts have to work together in exquisite cooperation for the most indifferent gaze to be maintained. The principles are pretty straightforward. The eye takes in and processes light in such a way as the visible surface of the scene being looked at is translated into a pattern of impulses in the optic nerve leaving the back of the eye which can be further processed in the brain and that visible surface reconstructed.

There are several fundamental problems with this seemingly straightforward story. The transformation of light energy into nerve impulses does not seem to explain the transformation of light into awareness of light and the patterns of light into patterns of appearances; in short, the process by which the scene is the seen. How the light gets into the brain via the eye seems pretty clear, but what is not at all clear is how the gaze then looks out, referring the light to the objects that are or were its source. And, finally, the sense of three-dimensional visual space, and indeed, the construction of a rich and complex visual world, seems an extraordinary, even inexplicable, achievement, given the two-dimensional surface of the retina which harvests the light. We shall address these fundamental questions – which apply to all sense experience, not just vision – in the next chapter. For the present, I want only to celebrate this marvellous structure, which, even if it is not a sufficient explanation of vision, is most definitely a necessary condition.

The light bouncing off the visible world enters the eye through the *cornea*, the clear, transparent portion of the coating that surrounds the eyeball. The cornea provides two-thirds of the eye's focusing power. It is extremely sensitive as anyone who has scratched it knows: the nerve endings are more densely packed there than anywhere else in the body. Although it is only about half a millimetre thick, it has five layers, several of which are charged with protecting the cornea from injury. The innermost layer, or endothelium, is only one cell thick; even so, it carries a heavy responsibility. It has to pump water from the cornea and keep it clear.

The amount of cornea that is available to let in light and meaning depends on how dilated the *iris* is, the adjustable diaphragm that gives the eyes their colour. (The opaque whites of the eyes are called the 'sclera'). The iris controls the light levels inside the eye, dilating in the dark and contracting in bright conditions. It is also under the influence of the emotions. In conditions of danger, the pupils – the round opening in the centre – dilate. Opiates make the pupils smaller and the pinpoint pupils of the heroin addict are a familiar sight in Accident and Emergency departments.

The iris has recently entered politics via the debate on identity cards and so-called 'biometrics' that provide unique identification of individuals wishing to exit or enter the country or to claim benefits. Iris scans analyse the features of more than 200 points of comparison, examining rings, furrows and freckles. This is more than enough to characterize uniquely every single person on the planet and banks may some day make iris scans a routine part of ATM transactions.

The general colour of our eyes contributes greatly to the immediate and enduring impression that we make. The rather tenuous connection between character and what is suggested by the indole monomer, a chemical that gives blue eyes their blueness, may be illustrated by the fact that 'A pair of blue eyes' can stand metonymically for the lovely innocence of a Thomas Hardy heroine, for the seductive charm of the rather less innocent Frank Sinatra, or for the sense of the far horizons stared into by the travel writer Bruce Chatwin. Many people commented on George Orwell's brilliant blue eyes, connecting it with his clearness of vision, his tough-minded, courageous decency and honesty. In fact George Orwell's public persona did not correspond to the duplicitous, private Eric Blair, tortured by sexual frustration and his unattractiveness to women.[2] And yet he had only one pair of eyes, whose colour was the result of their chemical composition.

The light admitted through the porthole of the iris passes through the crystalline lens. While the cornea does most of the focusing, it is the lens that does the fine-tuning, so that we can see near and distant objects with equal clarity. The lens is elastic and, without constraint, would tend towards a more spherical shape. An array of radial fibres – the zonule fibres – hold it stretched out, so that it is more disc-like. This is fine for distant vision, when the rays of light do not have to be bent too sharply to be brought into focus on the retina. When the object is close up, however, the lens needs to be rounder. The ciliary muscles contract and slacken the tension in the zonule fibres and the lens fattens.

The cornea, the iris and the lens are bathed in, and nourished by the 'aqueous humour' which occupies the anterior part of the eye. Its pressure is exquisitely regulated by balancing the 4cc a day secreted by the ciliary body that support the ciliary muscles with the amount that is drained from the eye via a spongelike three-dimensional network called the trabecular meshwork. When the outflow is blocked, the resultant rise in the intraocular pressure compresses the optic nerve and, if the hypertension is sustained, progressively destroys it. This can be an insidious, painless process but the outcome may be catastrophic. First peripheral vision is lost ('tunnel vision'), so the world is peered at rather than merely viewed and then blindness results. Undetected or inadequately treated glaucoma has made old age for many an endless night.

The space in the eye between the lens and the retina is called the vitreous body. This occupies most of the bulb of the eye and gives it its characteristic shape. It is filled with a clear, gelatinous material called the vitreous humour, enclosed in a delicate transparent membrane. For students of Shakespeare, the vitreous humour is inseparably associated with the blinding of Gloucester in *King Lear* (Act III, Scene VI):

> *Cornwall:*   Lest it see more, prevent it. Out, vile jelly!
> Where is thy lustre now?
> *Gloucester:*   All dark and comfortless.

The light arrives at its destination: the seven-layered retina, where it is transformed into nerve impulses. It passes through all the layers until it reaches the ganglion cells and is then processed by a very complex network of interconnected cells, ending with the optic nerves that carry the transduced light into the brain. It is complex in part because the retina has to operate under a very wide range of light conditions. The passage from dark night to bright sunlit noon amounts to a 100-million-fold change in ambient illumination. The eye's sensitivity to light is extraordinary. Under conditions of maximum dark adaptation, one quantum of light, absorbed by between five and fourteen light-sensitive rods is sufficient to generate a sensation of vision.[3] Without adaptation, normal sunlight would be blinding, and objects in the dusk invisible. Despite a huge research effort, the ability of the retina to cut its coat according to the available cloth remains shrouded in darkness.[4]

The retina contains many millions of specialized photoreceptor cells called cones and rods. The most sensitive part of the retina is the macula, where most of the cone cells are located. It is responsible for central vision and for seeing fine detail. Cone cells require a higher intensity of incident light and underpin colour vision. At the centre of the macula is the fovea, which has the highest concentration of cones and which is presently positioned by your eye movements to receive the light from this sentence. The periphery of the retina is where most of the rod cells are located. These require a lower light intensity but present the world in black and white. They are particularly important for night vision.

We are more alert to movement and change than to the unchanging. This has obvious adaptive benefits. We need to be awoken from the automatic pilot to the novel – to the edges of objects, to new things coming into view, to entities that are moving rather than still. The heightened sensitivity to change is evident throughout the nervous system but it is particularly beautifully exemplified in the retina, where there is a much greater response to a small spot or ring of light, or to a light-dark edge than to diffuse light. To make this possible, photoreceptors have a 'centre-surround' receptive field; that is to say, for maximum response, the centre must be mainly light and the surround mainly dark or vice versa. A given receptor will inhibit its neighbours when it is itself stimulated and they, in turn, will inhibit it. If Cell A is going to fire on all cylinders, it is best if Cell B, its neighbour, is not firing at the same time. This will obtain if the stimulus is a small spot, a thin ring, or an edge, of light.

The output from the retina is carried towards the brain by the optic nerve which itself punches a significant hole in the retina, where there is no light sensitivity: the literal, archetypal 'blind spot'. We are blind to our blind spot: the visual field does not seem to us to have two holes in it, corresponding to each of the optic nerves. This is a potent illustration of the way we tend to fill in what is not there; or, more generally, to see what we expect to see and draw from a scene the meaningful gist. It is also an immensely powerful metaphor, a physical reminder of the fact that at the heart of our viewpoint is something we do not see – or see only partially, distortedly or intermittently. That something is ourself.

Looking Deeper into Looking[5]

Let us consider what is special about vision compared with the other senses, so that we may speculate on what effect the dominance of vision may have on our relation to our material environment.

The first thing is that the object of sight is seen to be at a *visible* distance from ourselves. The object is 'over there', distinct from my eyes, indeed from my head. Its independence of my head is underlined by the fact that movement of my head and movement of the object occur independently of one another: the head may move without the object moving and vice versa.[6] Importantly, we do not merely infer that this is the case: we *see* that this is the case. Not only the object, but also the distance that separates me from it, is visible.

There is nothing like this in any of the other senses. We do not smell the distance between a smell and ourselves; indeed, there is no such olfactory distance, though there may be a spatial distance between us and the smelly object. Nor do we smell the distance between smells or between smelt objects. (Actually, smell per se does not give us objects, in an important sense which we shall examine presently.) We do not touch the distance between ourselves and the object we are touching. While we may be considered to experience this 'tactile distance' indirectly through the experience of our reaching arms, it is not itself made of tangibles; for intervening tangibles would be obstructions. Similarly, although we may conclude that a sound is coming from a remote source 'over there'- as most sounds do – we do not *hear* the distance of the source of the sound from us. After all, a faint sound could be one that is relatively quiet at its nearby source or relatively loud at its distant source. To put this another way: the location of the source of a noisy object is an inference in the way that the location of a visible object is not. As the philosopher P. F. Strawson put it, sounds 'have no intrinsic spatial characteristics'.[7]

Secondly, we *see* that there is more to seen objects than what we are currently sensing. This is the respect in which sight, but not smell, gives us 'transcendent' objects. Doesn't touch give us objects? When I handle something, when I lift it up, or press it, or bite it, do I not get a sense of 'more to come', of it being more than the surface I am touching, through its weight and its resistance to my fingers or teeth? Yes, but not in the way that I do when I am looking at something. This warrants further exploration: it is fundamental to the nature of the human world.

When I see an object, I see that the visible surface conceals invisible depths or surfaces not yet revealed to me. The front of the object hides the back, the upper part the lower, the outside the inside. The visual field, in short, is made of opaque objects, of entities that are for the greater part unseen and may also occlude other objects. If objects were entirely transparent, they would be invisible: edgeless, shadowless and visually contentless. Visual revelation, in short, depends upon explicit concealment. The visibly hidden is the necessary condition of the visible. This applies not only within the visual field but also around it. The visual field is limited by an horizon. This line, part real, part conceptual, joins all the local boundaries between the visible and the invisible, and is itself visible. And there is a further large and continuous mode of the invisible – that which is outside of the limit of our visual field – at the back of our head.

Something else follows from all this: the visible is continuous. There is a visual *field* in a way in which there is not an auditory *field*, an olfactory field or a tactile field. The space between sounds is not itself made of sound: as we already noted, sounds have no intrinsic spatial characteristics. We may, as it were, fill in the dots between sounds we locate at different points in space; but that which fills the gaps between the dots is not itself made of sounds. Smells may be all pervasive but they are not explicitly continuous in space and time. And while it is possible to trace a continuum of material substances with touch, this continuum is not available all at once. Touch is confined at any given time to the touching surface of the body. The untouched – about-to-be-touched or having-been-touched – surfaces exist only as possible sources of touch. What a difference from vision, where all the visibilia in a visual field exist in an instantaneous relationship with one another and form a continuum of the visible and the visibly invisible, of what we can see and what we can see that we cannot see.

All of this gives me a sense of the object in itself, as being distinct from myself and it belongs to a visual field in which it is related to other objects.[8] The actual distance between our eyes, ourselves, and the objects of our vision is very important. The seer is not immersed in that which it sees. Vision lifts us up above the sea of things. We do not have to mingle with the seen to be aware of it. Our elevated, seeing head, from which, uniquely among the higher mammals, we humans derive most of our sensory information, is a constant reminder of our distance from that of which we are aware. The light by which we see what we see is more refined than the smells that pervade our nostrils or the collisions that mediate touch. The animal sniffing its way around the world is doubly in the muck with its nose down among the *merde* of the natural world.

Uniquely among animals, humans have a sustained and complex

sense of themselves, rooted initially in their appropriation of their bodies as their own, as themselves. Humans are *embodied subjects*. I have a sense 'That I am this' where 'This' in the first instance is my own body. This makes my visual field into something which has me as its explicit centre. I stick out, like a fat thumb. I am related to its objects, and those objects are related to me. The relationship is not, of course, symmetrical. While the objects exist in my visual field, I do not (with exceptions to which we shall come shortly) exist in theirs. I see the world around me as being *around me*; and, located in that world, I see *that* I am seeing its objects *around me*. What is more, I see that I see them from a particular angle, in a particular light, and so on. And I consequently see that there is more to see – from different angles, in different lights.

I experience myself as a viewpoint in the view. A viewpoint is only one of many possible viewpoints; and I see that I am seeing from only one possible viewpoint. Other seeings, other visual fields, are possible – available to myself in another position, in a different light, at a different time; or to other people. We shall come to other people presently and for the moment focus on a couple of fundamental cognitive consequences of my sense of being at the centre of a visual field in which there is visibly more to come.

The first is the passage from passive, or random, experience to active inquiry. We may postulate a gradient of inquiry, extending from dazed looking, to gawping, watching, to staring, to peering, to scrutinizing, to observing and finally to investigating. This passage is facilitated by the fact that the seeing human being is raised above the objects, the world, of which it is the centre. It is elevated above the muck, in the sense of being uncoupled from it – being linked to it not by direct contact as in touch or tastes but by the mediation of light. This elevation is literally manifested in the elevation of the head, its being raised to the top of the upright body. The combination of a sense of 'more to come' and of being distant from, outside of, other than, that which one experiences, takes us down a path leading from the merely sentient creature dissolved in its experiences to one who seeks out experiences in pursuit of more experiences and knowledge – for itself.

It is highly appropriate, therefore, that most of our metaphors for knowledge and, more broadly, cognition are based on the visual sense. It is difficult to imagine how without vision (and one or two other things, such as fully developed hands with an opposable thumb, as I have argued in *The Hand. A Philosophical Inquiry into Human Being*)[9] the experiencing beast could have been transformed into the Knowing Animal, that encounters objects of *knowledge*. We speak of 'envisioning' or 'imagining' things, and of those who have 'vision' or are 'visionary'. We describe intelligent people as 'clear-sighted', of having 'insight', of seeing things in a certain way, of 'speculating', and of our sense of the world around us as 'a world view' or 'world picture'. Indeed, the very feeling that we are located in a continuum – the, or a, 'world' – is the product of the sight which synthesizes what is before us into occupants of a (visual) field. The Greeks, who were the first humans to theorize in a systematic way, to find theories to underpin observations, and even to have a theory of theory, derived their notion of theory from that of *theoria*, which literally meant 'a looking' or a 'viewing'. They valued the attitude of the disinterested spectator, in whom the visual distance, or uncoupling, was taken further to a distancing from any kind of engagement, even beyond sensory immersion, to a 'beholding', or pure contemplation of what was there.

I do not for a moment want to suggest that blind people, even people who are blind from birth, have an entirely different world picture from those who are not blind; that they do not have an intuition of objects separate from themselves or of a continuous sensory field that adds up to a world. Blind people have grown up in a world dominated and defined by thousands of generations of the sighted. They have swallowed human epistemology and ontology with their mother's milk, and they are nurtured during cognitive growth. Like the rest of us, blind people take most of their world 'off the shelf'; they simply take more of it. And that off-the-shelf world is overwhelmingly mediated by words. Such words as 'over there', 'a few yards to the left', 'a couple of miles away' endlessly reiterate the spatial metaphysics of everyday life, a life in which blind people participate as completely as the sighted. The notion of the space as a continuum, literally made visible in the visual field, goes through every aspect of our practical engagement with the world and everything we say about it, irrespective of whether we are blind or sighted. What is visible to sighted people is still explicit in the world picture of people who are blind.

In this discussion of the metaphysical significance of vision, I have so far talked only about the early stages of inquiring consciousness, and have said nothing about language, not to speak of writing, which, as we discussed in previous chapters, have vastly extended the universe of knowledge. It is worth exploring for a moment a profound link between the visual sense and language. Language is not merely a set of cries wired into the material or natural world in virtue of being responses to stimuli, but is a conscious assertion '*That* something is the case'. It confronts objects in the world, acknowledges their independent existence, and recognizes their public reality – that they are available to the pooled consciousness of the collective to which we belong. It is not, I hope, too fanciful to connect this with vision and the mysterious alchemy by which the input of light into the eyes is turned back into a gaze that alights on the source of the light itself, which is at a visible distance from the eyes.

This turning back of consciousness, of experience, on to its own putative source, cannot be explained by neurophilosophers.[10] It is, however, the heart of the process by which scattered light, experienced by a human subject, is turned into *knowledge* of the separate existence of the object that scattered it. This is massively elaborated in speaking subjects whose headwinds assert of what is there *that* it is there. The wired-in animal is merely part of a causal nexus and is not aware of itself as being located in, confronting, that causal nexus as something other than it. The seeing human animal is of course located in the causal net, but explicitly so, and is aware of standing in relation to it and of the latter as an explicit object of inquiry: that which is visible explicitly conceals what is invisible. The passage from this primordial explicitness to language and the assertion 'That x is the case', 'That such and such exists' took place much later, as recently perhaps (as we have noted) as a mere 40,000 years ago. The crucial parting of the ways between humans who do and other higher primates who do not have a sense of objects separate from themselves, took place hundreds of thousands of years, perhaps several million years, before language came on the scene and the scene was so dramatically changed.

The benefits of this visual (and subsequently verbal) awakening out of organic immersion are, of course, huge. The elevated animal interacts with nature on different, more favourable terms – though the final outcome (death) is unchanged. He lives a safer, more comfortable and, in some respects, incalculably richer life. But not everything is gain.

Both the loss and the gain, arising from the distance opened up between nature and ourselves, were identified very early by Aristotle who must, surely, have been thinking about vision when he defined perception as 'the capacity to receive the sensible forms without the matter'.[11] When I look at that object over there, its form enters me, but not the material of which it is composed. This is the point at which the gap opens up between the human being and the natural world; at which sense experience passes over into the knowledge '*that* such and such exists or is the case'.

The massive elaboration and enfolding of this 'that' – in suppo-sitions, propositions, facts, theories, ideas, all the manifold manifestations of discourse – creates a human world, a culture that separates us from nature and, in some respects, from the bodies from which we have arisen. Thus are our heads distanced from themselves. We are consequently profoundly self-divided creatures, experiencing a gap between what we immediately experience and what we know and take account of in our lives. While this confers upon us the privilege of *leading* our lives rather than merely living them, we are deeply unhappy, eaten away from within by possi-bilities, by ideas no experience can correspond to, by a regretted past, and a feared future, separated by an elusive, unsatisfactory present.[12]

Let us turn our gaze nearer to its origin. I am aware of myself as a viewpoint within my visual field, and sometimes, as a kind of emplacement for my viewpoint, with my body, and more particu-larly my head, as the material basis and immediate surrounds of my gaze. This is especially likely if I have difficulty seeing or vision is painful. The screwed-up or hung-over gaze peering through holes cut in a burqa of malaise, the dazzled gaze shaded by the hand, the gaze squinnying through cheap spectacles – these are aware of themselves as the condition of the scene being seen and are aware also of themselves as being located in the seen. This awareness is present, though more dimly, even under normal con-ditions: I keep glimpsing the corner of my eye as the material presence of the corner of the world from which I am seeing that world. Pause for a moment and attend to your awareness of the immediate bodily surroundings of your gaze that make your eyes present to you, so that your viewpoint is literally located. Not only is the viewpoint specifically located; it is fattened from a notional point to something more extensive – if blurred, variable and evanescent. This is particularly evident when I blink, or my gaze is misty with tears, or my eye is sore; then I am reminded of the smearing of my point of view over the apparatus that makes it possible.

The fact that I am an impure viewpoint incompletely effaced before its view, rather than a pure metaphysical subject, or a geometrical point at the centre of egocentric space, makes explicit my distinct existence as a subject in the world, confronting objects that I am not. This impurity underlines our anomalous condition as embodied human subjects half-awoken and quarter-liberated from our organic origins. Although the curdled blobs that are our individual viewpoints have joined forces so effectively with other curdled blobs to create collective viewpoints whose views look far beyond the immemorial horizons of sense experience, and reach into the distant past and the remote future and the interior of stars, they have their troubles. Aware of being located in a boundless world, we are not sure of our place in that world. We are distanced from it: Kafka, with the capacity of tortured genius to experience the metaphysical conditions of human existence as a personal problem, spoke of being 'separated from all things by a hollow space'.[13] The objectification of the world around us is, as Heidegger said, a kind of '*de*-experiencing' : our unity with the world is broken up and we become subjects confronted with alien objects.[14] And, finally, the great spaces envisaged in our knowledge seem to diminish us: we seem increasingly small in an expanding cognitive universe.

I don't want to end this section on too bleak a note. It is a good moment to remind ourselves of the pleasures of the recreational gaze; of looking for the joy of looking, of the delicious condition of being *lichtbetrunken*. And how rich and folded these pleasures are. I visit an art gallery to enjoy the colours and forms on the canvas; the various tensions between surface and depth, between the material of the paint and those things it represents; to admire the genius of the craftsman; to read the story the canvas tells about an event, a face, a city, or mythical past; to learn about the development of an artist or of art; to accompany someone, to impress them. I wander the streets of a city to participate in the many different delights of the spectator, of the stroller, of the onlooker, of the loafer, of the *flâneur*, of the nosy parker; of the traveller who wants to broaden that interesting construct he calls his 'mind'. On a country walk, we transport our heads through nature, trawling for views. On a mountain climb we pause for breath, ocular heaven poised on physiological hell... We look around the great spaces of the world and expose foreign parts to our partially comprehending gaze. And how much care we take over these occasions: joining the rambling club, preparing sandwiches (alas to be eaten as we take in the view through a rained-upon windscreen); buying tickets, guides, currency, and mugging up language, culture, manners, architecture, restaurants...

Alas, it is not all sweetness and light even here. The experiences we seek out do not always correspond to the ideas that led us to seek them out. Experiences sought for their own sake may seem less clear, less tidy, less purely themselves, than memory and anticipation lead us to expect. We are brought up against the gap between what we know or imagine and what we experience, between what we had in mind and what lies before us. The game on the beach doesn't quite match the idea of the game on the beach. The dip in the sea seems to divide into preparation and recovery and not much else between. A holiday unpacks itself into an entanglement of arrangements, of plans, of plans about plans, and by the time we have shed the preoccupations we brought with us, it is time to go home. We try all sorts of ways of perfecting the experience. Digital cameras enable us to capture the view, or to feel that we are capturing it, so that we are distracted from our inability to lose ourself into it. The quest for perfected experiences is never satisfied: we are driven to consume more and more, to buy more and more, in a vain endeavour to find a sight that corresponds with what we had in our mind's eye.[15]

Being Watched

> The ear receives, the eye looks...one can terrify with
> one's eyes not with one's ear or nose...
>
> Ludwig Wittgenstein, Zettel

The gazer, aware of his presence in his own visual field, aware of his own opacity and hence visibility, knows that he is also gazed upon.

While animals may return our gaze, it is the gazes of our fellow humans that touch us most deeply. When we are the terrified prey of an animal predator fixing us with the pure malignity of its stare, when our lives are at stake, we may feel powerless, but the predator is still not our equal. It is halfway between an avalanche and a sentient being: though full of menace, it does not judge us. Our fellow human beings do judge us: their claws enter our souls, not our bodies.

Let us get down to basics. The thinker whose gaze at the gaze has been most influential was a wall-eyed, myopic, rather short French philosopher, the most famous since Voltaire: Jean-Paul Sartre. He stared at the gaze through pebble lenses, tinted by the philosophy of Hegel, and came to some rather bleak conclusions.

For Hegel, the deepest hunger of mankind was for acknowledgement by other human beings. 'Acknowledgement', as Heidegger says, 'lets that towards which it goes come towards it'.[16] Every self-consciousness, Hegel said, can be satisfied only by another self-consciousness. This is what makes us different from the animals. Whereas animals hunger for food and thirst for water, humans – once they are fed and watered and are warm and pain-free – hunger and thirst for acknowledgement by others. They will go crazy for love. This is both good and bad. It is good, since wanting to be loved may make us behave well towards others: it is the foundation of morality. But it is also bad, not only because it encourages us to be deceitful, so that we pursue our own ends under the guise of caring for others, but also because it is the basis of a power struggle. Each would like the other to acknowledge herself while not necessarily acknowledging the other. Each would be Master rather than Slave. It is this aspect that attracted Sartre.[17]

Since Sartre, everyone knows what happens when A looks at B: B is reduced to an object in A's visual field. Then B looks at A and tries to turn A into an object. And so the tussle begins, in which A tries to locate B in her world and B tries to locate A in his world. Fortunately, the interactions between human beings are not as simple or naked. Even if this conflict were the primary relationship, it is elaborated, qualified and modified by the lives and agendas through which it has to be enacted. Our encounters are rarely one to one. Even one to one relationships are between individuals who are steeped in histories of other relationships (which may be more important) and have other preoccupations. So we can work together in teams, or exist happily or indifferently side by side, despite being in others' visual fields. Indeed, society can be seen as an infinite nexus of shifting teams. What is more, we enter each others' visual fields sheathed in a many-layered carapace of a self that distances us from our bodies and remain unexposed by, or unengaged with, the others' gazes.

There are, however, extreme situations – torture, the supreme moments of sexual love or hatred – where the question of who is located in whose world is reduced to a naked struggle between two people reduced to their visible presence. But these are not typical of everyday life, where our encounters with others are diluted by outer or inner crowds, muted by preoccupations, and rendered less conflictual by a shared agenda, tasks, goals and responsibilities. While the primary conflict may erupt at any time, for the most part it does not.

It is easy to see why the gaze makes an attractive metaphor for human relationships and, in the hands of a master of dramatization such as Sartre, a persuasive one. The Peeping Tom, who possesses the world he looks at until he realizes that he has been caught peeping, so that his own gaze is gazed upon and he is himself possessed; and the solitary walker in the park who sees that the park is not his alone when he catches sight of another – these are compelling instances. By locating one's self in their world, the other's gaze seems to displace one from the centre of the world: because I, too, am surveyed, I am not 'Lord of all I survey'. From being the judge, classifier and consumer of what is there, I become the judged, the classified, the consumed; I fall from transparent, invisible subject to opaque object visible to another subject. The myth of Medusa who, catching sight of her own gaze reflected in a soldier's golden shield, is herself turned to stone, is an intriguing variation on the theme of the fall of the gazer, when he is gazed upon.

An attractive metaphor, yes, but not an accurate account of how it is. The Eden in which the looker was the centre of the world upon which he looked never was. Once we are explicitly located in a world, we are not alone. We *begin* living in relation to others – and often to many others: the looker is most often an onlooker. Fundamental to vision is the sense of the world as something that transcends us. As a consequence, we as often feel that we are as much at the periphery, as at the centre, of 'our' world – even if this sense of being on the periphery is itself located at some kind of centre. This is never more manifestly true than when we are relying on someone else to tell us what is there. The information that we need underlines how the world lies largely beyond us.

This is evident at the most primitive level in pointing.[18] When something is pointed out to me, I am required to assume the viewpoint of the pointer-out in order to see what is being pointed out. In order to receive the information you have given me, I have to see what you mean; and this requires my 'seeing where you are coming from' – in other words imagining into your world, into your point of view. I have to put myself in your shoes.

This is closer to typical human interactions than is Sartre's imaginary tussle between self-consciousnesses each trying to reduce the other to an object in its own world. The medium of human life is a world of knowledge that is established, maintained and propagated in and through a community of minds somehow shaking down together. At the heart of the collectivization of consciousness, of the shared world in which we live, is the intuition of other viewpoints than our own. At the root of this is the acceptance of the objectively true fact that my viewpoint is only *one* viewpoint. While others are a permanent challenge to our sovereignty, the world in which we would be sovereign is the joint product of this understanding.

Nevertheless, the power of the gaze and the profound discomfort attendant on being watched cannot be underestimated. The light from my body enters your eyes and comes out stained with a judgement rooted in yourself and your world. The judgement may be concealed or may be expressed, and it may be approving or disapproving, but its potency cannot be denied. The pain of looking foolish, of being exposed to the eyes of others, is as real as toothache and may be even less bearable. The light in which your gaze bathes me comes from an individual but it carries with it a greater authority. While your gaze originates from within you – and so lies beyond my reach fully to know and certainly to amend – the hidden world of judgements and thoughts to which it belongs in turn draws upon the collective which is even further beyond my ability to control. There are individuals who are supreme masters of the arch look. As you talk to them, you feel judged, questioned, undermined. Everything you say turns into 'a likely tale' and even wishing them 'Good morning' seems a rather odd and suspicious thing to have done.

Reciprocating a gaze is an important part of the ballet of acknowledgement and non-acknowledgement that flavours our interactions with others – even where (as in the vast majority of occasions) these fall short of Sartre's existential fight to the death. When I look at your leg, it does not look back at me. When I look at your eyes, they do look back at me. Indeed, if your eyes are not averted, I say that I am looking into your gaze. I feel that I am looking at, or into, *you* – even though I cannot penetrate past the lens, and the humours, and the retina and the optic nerve and the brain to your awareness of me – because your eyes are the point of origin where you feel your view is rooted. My gaze touches the place where you feel yourself to be. In your gaze, my sense of self is heightened, even though I may feel that that self is being traduced; when the gaze is directed explicitly towards that part of my body that is most my self, this sense of self is heightened further. We feel, as the ordinary saying is, 'self-conscious'.

It is hardly surprising that the rules governing the exchange of gazes are so complicated. It is the potential symmetry of the gaze, in a world of asymmetrical relationships, which makes for complexity. In some cultures, or in certain circumstances, not meeting the other's gaze is a prescribed courtesy, evidence that one does not have the worth of the other. The courtier does not reciprocate the gaze of the king, the woman that of the man. Modesty with respect to social standing or sexuality (and in the case of women in most cultures, both) demands the lowered gaze. In an unjust and asymmetrical world, the bold stare is the sign of those who, not in possession of the world, behave as if they are: the insubordinate subordinate, the insolent youth, the 'hussy'.

Under other circumstances, the reverse is the case. The one who looks expects the returned gaze of the one who is looked at, as evidence that the other's attention is engaged or that their own presence, even their humanity, is being acknowledged. 'Look at me when I am talking you!' is the familiar cry of the exasperated parent, trying to hook the undisciplined child's truant attention. The response may be a defiant gaze that tries to 'outstare'. The failure to make eye contact, to interact with the eyes of one's interlocutor, is one of the primary manifestations of autism. This is a condition in which an individual, who lacks an integrated sense of self, also has a diminished awareness of the selfhood of others.[19]

We are offended by professionals who do not make eye contact. They may deal with our problems without fully acknowledging our humanity. This is particularly important in medicine, where we feel deeply unhappy if our worries or suffering are reduced to an instance of a case. A doctor who looks us in the eye says we are persons rather than simply the other pole of a transaction. Nowadays, doctors undergoing their regular appraisal are often videoed interviewing patients and are scored positively for eye contact. The situation, however, is often more complicated than can be catered for in general prescriptions. There are times when the patient's shyness should make one more circumspect. The doctor's gaze, however, is now gathered up in a wider gaze and common sense and true sensitivity are struggling for survival in the face of rules and regulations.

It is because the gaze potentially acknowledges the humanity – the equivalence-to-me – of the other, that it is so subversive. It erodes the carapace of the CV, the established order. In *War and Peace*, Pierre Bezukhov has been taken prisoner of war. He is brought before the notoriously savage Marshal Davout (who used to rip off soldiers' moustaches at parades of inspection). He will probably be executed:

Davout looked up again and stared closely at Pierre. For several seconds they looked at one another, and it was this look that saved Pierre. The business of staring at each other took them beyond the realm of warfare and courtrooms; they were two human beings and there was a bond between them. There was a single instant that involved an infinite sharing of experience in which they knew they were both children of humanity, and they were brothers.[20]

Not all gazes are so affirmative of the other's humanity. The hostile gaze has many variants. There is the gaze of contempt that looks down its nose. There is the inspecting gaze that examines one as a specimen. This has a long history and even has archaic types – as the gaze that peers at its object through a lorgnette. There is the look of indifference.

The power of the gaze is testified to in the literature of love. 'Love in her eyes sits playing and sheds delicious death'.[21] An entire movement in poetry – *Il dolce stil nuove* – seems to focus on the icy terror and even mortal illness that steals over the soul when the mistress looks at the hapless poet.

> And I, who gazed fixedly upon her
> Am in danger of losing my life
> Because I received such a wound from one
> whom I saw within her eyes.[22]

There is the fierce gaze of the all-powerful male. And the calculated, easy gaze of the seducer. Natasha in *War and Peace*, hardly less innocent than a child, is doomed to betray her fiancé Prince Andrei, after only a few minutes' exposure to Anatole Kuragin's gaze:

But when she looked into his eyes she was shocked to realize the usual barrier of modesty that existed between her and other men was no longer there between the two of them. It had taken her five minutes for her to feel terribly close to this man, and she scarcely knew what was happening to her.[23]

The power of the gaze is dramatically attested to in Andalusian culture. According to John Richardson, the cult of the *mirada fuerte* ('strong gazing') lay at the heart of Picasso's genius and his legendary powers of seduction.[24] His huge eyes and fierce stare possessed the world he looked at, so that he could transform it and remake it as his own thing.

> When the Andalusian fixes a thing with a stare, he grasps it. His eyes are fingers holding and probing…The *mirada fuerte* has elements of curiosity, hostility…and envy. But the sexual element is present also…The light of the eyes is highly erotic.[25]

The gaze, modulated to a glare, is a weapon. There is the stare that will not be outstared. A direct, primordial assertion of hostile power, it challenges the other not to blink first, and matches the other's response volt for volt. It is possible to 'look daggers'. The gazed upon may be frozen in terror, as malign intent is transmitted from the eyes. The victim cedes the initiative and waits dry-mouthed for what is going to happen next. The violence of the gaze fries the consciousness of its victim. Most powerful in this respect is the gaze that suddenly appears from nowhere: the face at the window. Think of the terrifying apparition of Peter Quint outside the dining-room window in *The Turn of the Screw*, a story in which the language of the gaze, of vision, is all-pervasive.

There is power in the gaze and there is power in withholding the gaze. You look at me and I bask in your gaze but do not return it. I refuse your request for attention and enjoy the attention paid to me. Or I sneak a look at you but do not let you catch my eye or even catch me looking. I want to help myself to your appearance – to look that bosom, to see those tears – without your knowing it, or letting you know that I know you are aware of my doing so. I steal information about you without your being informed of the fact. If knowledge is power, to be known – and to be known without knowing that one is known, or at least in a way that one has not specifically chosen or given permission for – is a form of powerlessness.

One of the most popular ways of looking without being seen, of gazing without returning the gaze, is to wear dark glasses. This is why they are so favoured by what my parents would call 'criminal types', even in darkness. They would rather rob, than be given, another's visual appearance. It is certainly disconcerting to talk to someone in shades, as I remember when I was a seemingly permanently cross young doctor in Nigeria. Time and again, the laboratory specimens would go astray. The young man who was responsible for portering enjoyed socializing on the journey from the ward to the laboratory. I used to rail at him. I gave it up when I saw my own image – white and angry – reflected in his mirror lenses as he smiled at me.

There is a sense in which all gazes are unreciprocated, because we do not know what soil they grow from, what is withheld, what thoughts the gazer has. We were so witty that evening and it looked as if we had made a big impact on the girl we fancied. Later we learn that the enigmatic smile that accompanied her seemingly dazzled gaze was a private joke prompted by the bit of egg on our face, that she didn't have the heart to tell us about. The most candid gaze from the most open face, which seems to expose itself at least as much as it exposes those on whom it looks, has a hidden history. Even when we look in the eyes of one closest to us, who returns our gaze with love and equality, and we have the sense of being seen for what we are by one who is seen for what she is, there remains a great world of darkness unilluminated by these interlocking eye lights.

There is, of course, one gaze which gets us right and sees us just as we are. This is the sight of God – the gaze in which all judges and authorities converge, all signs are cashed, and the self freezes from fugitive impressions to an eternal soul eternally judged. God has complete knowledge not available to human gazers: such a gaze is to be feared. Fortunately it cannot be imagined either; for God's viewpoint is a view without a point – the view from that everywhere which is nowhere, that is the final truth of things.[26] It is as impossible as an impersonal view of our person.

There is no place, therefore, where the truth behind our gazes is available. We do not even know it ourselves. This does not inhibit us from interpreting gazes. We believe that we can read people's characters from their eyes. We are entranced by the intense cobalt blue of the iris and think that this corresponds to a cleanness of spirit. We talk about intelligent and beautiful gazes. They have little relationship to cognitive or spiritual reality.

We are attuned to looking at and reading gazes. We can detect an eye movement of less than two millimetres when standing up to a metre away from the face, and this skill is developed very early in life. As Chris Frith has said:

> This sensitivity to eye movements allows us to take the first step into someone else's mental world. From the position of a person's eyes we can tell very accurately where they are looking. And if we know where people are looking, then we can discover what they are interested in.[27]

The eyes can be used in very subtle and complicated ways as instruments of deliberate communication. I can dilate my eyes – more strictly my orbital fissure, put into upper case by my eyebrows – as a silent warning of someone's presence, that something is behind you, that you are being tactless. I can raise the font of my gaze by turning it into an otherwise unexplained intense stare. By this means, I communicate something to you without betraying the fact that I am communicating it. Alternatively I may use my eyes as a pointer: I look in a certain direction and, as with literal pointing, you are instructed to look in that direction so that you will see the thing I am seeing which I may, for all sorts of reasons, wish you to know about. The gaze, in short, is a signal of itself.

In view of the intense significance of the gaze, it may seem extraordinary that we can withstand the gaze of so many when we give a lecture, or address a crowd. Perhaps we can handle it because the crowd is agglutinated into a single gaze and each contributing gaze is diminished by all the others, or we pick out one or two faces from the crowd, who can be addressed as individuals. (Middle of the third row the experts tell us.) Indeed, we can turn the tables, as teachers do, by singling out individuals to answer questions. Then it is those others who feel outnumbered, as all gazes are upon them.

Let me end with a toast: the Viking 'Skol', uttered as the warriors raise the empty skulls of their enemies from which they drink wine. As they do so, each has to make eye contact with all the others, to test them for duplicity. The skull is a good reminder of the end of gazing, when the living daylights go out and the looker can return no more looks.

chapter nine
The Sensory Room

The Mystery of the Head-Room

The head is a listening post, harvesting music and noise, meaning-ful sounds and meaningless racket; a sniffer of scents, relishing and loathing what it finds; a taster of tasty and distasteful tastes; and an attentive student of itself. The fact that it does or suffers all these things at once directs us to our first port of call. For it should be chaos in the sensory room – consciousness should be conscious mess – and, so far as I can tell, it is not.

At any given time, I am hearing sounds, smelling smells, tasting tastes, being haunted by the proprioceptive ghost of my head, as well as seeing a world of visibilia and attending to the things I am thinking, the images I am imagining and the memories I am remembering. Some people refuse to find this at all problematic. They argue that the sights, the sounds, the smells, the tastes, and the sensations telling me that my head is heavy, warm, tingling, aching, or simply there, each have their own 'input channels'. Tickled up by incident energy, these channels deliver nerve impulses to neatly compartmentalized areas of the brain, so that experiences are not muddled with one another.

There are, however, numerous problems with this neat model,

not the least being that the things that are kept tidily apart have somehow to come together in order to be experienced as part of the same moment of consciousness, to belong to the same 'conscious field'.[1] Let me say a little about this, before we return to more homely aspects of the head as sensory room, focusing for a moment on the inadequacy of the neural account of consciousness.

As you sit reading this, you are perhaps aware of an itch on your cheek, belonging to the same moment of consciousness as the sounds you overhear – birdsong, music, an argument – the pressure of the chair on your buttocks, the sensations of your fingers gripping the book, the sight of the sentences on the page, the ghost-sound of your commentary on what you are reading and so on. How do all these different things converge in the same moment of consciousness without losing their separate identity? How do they merge or become integrated into a whole without ending up as mush? If this seems difficult, it is even less easy to explain the coherence of conscious activity over time necessary for my understanding of what it is that is going on around me. Making sense of the sounds, the sights and so on that I am currently experiencing requires that they should be pervaded by memories. Only in this way will there be a smooth passage from a comprehensible past to a comprehensible future via the present. And yet these memories of the past, and the anticipations of the future they shape, have to be kept distinct from the present.

It is worth dwelling on this and reminding ourselves why we need to come together as well as to remain tidily apart. Consider the long-range, explicit internal connectedness of consciousness that is necessary to be a responsible agent who is able to operate effectively in our complicated world. The time frame of such an agent is huge and multidimensional. Let me illustrate this with something that, as an academic, I do rather frequently – giving lectures abroad. A little while back, it was in Hong Kong. This was the fulfilment of a commitment undertaken many months ago. Any such commitment brings together a multidimensional lace-work of conscious moments; for example those in which I discussed the title of the talk in correspondence with my host, those widely scattered occasions in which I undertook its prepara-tion and, finally, those moments in which I deployed all sorts of implicit and explicit knowledge in order to find my way via foot and taxi and plane to Hong Kong at the right time in the right place, while in the grip of a thousand other preoccupations of an academic clinician, and all the while floating in a sea of sense data, as experiences flooded incessantly into my sensory room.

Now, however much my audience may have regretted my suc-ceeding in pulling off this feat of getting to speak at the appointed time several months after the idea was first floated, it is a remark-able tribute to the inexpressibly elaborate inner organization of my life and its extendedness across time. And this is where neuro-science lets us down. Somehow, bursts of electricity in the wetware of the brain don't seem adequate to the exquisitely structured mind that I, and you, have. Breaking up consciousness into modules mapped on to discrete compartments of the brain doesn't ease the problem – and this is not just because the number of modules required seems to be infinite. While breaking the mind into modules serves the purpose of keeping things tidily apart, it frustrates the need to bring them together in the moment of con-sciousness.

The troubles go deeper than this. If you think of all the things that would have to be going on in my brain in order to sustain my behaviour when I am giving a lecture, you could be forgiven for entertaining the image, based upon conventional neuroscience, of a vast number of overlapping electronic microcircuits supporting a huge ensemble of different functions. Now it is difficult to see how they could not interfere with one another. Let me give you an analogy. Think of a million sets of ripples in a pond created by the impact of a dense shower of hail, compounded by all sorts of internal sources of ripples. How are we to explain how each ripple or set of ripples, such as those supposedly corresponding to my plan to talk in Hong Kong, could retain its separate identity? It seems impossible, doesn't it? It seems even less possible if we remember that, ultimately, the nervous system has to allow everything to merge in the moment of present consciousness, steeped in meaning, but retaining its relation to a highly structured near and distant past and reaching into an equally structured future of expectation, responsibility, timetable, ambition and life plan. This moment (unlike the present moment of a computer, even a Cray super computer) has to bring everything together, so that I know where (in the widest sense) and who (in the deepest sense) I am.

What makes bringing everything together difficult to understand in neurological terms is that lots of things must still be kept apart. For while the events in the brain have to add up to some kind of unity, the brain must at the very same time keep vast numbers of projects, actions, micro-projects, micro-actions, distinct. Moreover, to make things even more difficult, those distinct projects must connect with a thousand others, if only because each provides the others' framework of possibility. My being in Hong Kong for the talk explained my refusing other invitations, rearranging the day so I arrived there on time, my being particularly concerned to keep my distance from someone who had a cold a couple of days before. The distinctiveness of the patterns of ripples has to be retained although the patterns have to be open to one another.

It gets worse. Moment-to-moment consciousness has to retain a global openness in order that I can enact my planned activities in a sea of unplanned events: so that, for example, I avoid the cyclist who might have killed me as I crossed the road to the lecture theatre, or I take account of the fourth step outside the hall as I am greeted by another of my hosts, on my way to accomplishing the timetabled task. We tend to overlook the complexity of the most ordinary aspects of our lives when we think about the supposed neurophysiological basis of consciousness. Neuromythology – which claims that neuroscience can explain far more than it can – seems halfway plausible only if it is predicated upon a desperately impoverished account of our many-layered, multi-agendaed, infinitely folded but wonderfully structured and organized selves.

We could summarize the problem very simply as follows. If we try to make sense of the unity of consciousness by adopting a holistic account of the brain, we encounter insuperable difficulties in explaining why so many different things which have to be kept apart are kept apart and do not collapse into consciousness-mush or delirium. If we try to address the problem of the multiplicity of distinct elements of our conscious lives by adopting a localizationistic or modular account of the brain or mind, we encounter equally insuperable difficulties in explaining how everything comes together sufficiently for us to live active, coherent lives.

And they *are* active – I am a doer of things rather than merely being the seat of happenings. Activity may be at many levels – as when writing this book, I also listen out for the phone or, fed up with writing, pause actively to attend to some music I have half-heard emanating from downstairs. And they are coherent: what I did last Tuesday or last year will make sense to me when I recall, even defend, or draw upon it, next Tuesday or next year.

The question of unity and control amid diversity and passive openness to the continual rain of the half-expected unexpected picks out a deeper problem: that of accounting for the fact that there is such a thing as 'the first person' – the I, here, now – to which all this variety is ultimately referred. Without such a unifying I, the brain or mind would simply be a colloidal suspension of unhaunted modules – which is how the cognitive scientist seems to present it. That is why many neuroscientists deny that there is such a thing as a self. If they can't find it or conceive of it in neurological terms, it can't exist.

The notion of the first person not only highlights the unity of consciousness necessary for me to act as a responsible agent in a bewildering world, it also opens on to a deeper issue: the origin of the sense of me, here, now, the suffering agent, the responsible creature, who is a *viewpoint* on the world. It is no use saying that my brain gives me the sense of me here now because it *is* my brain and it is here and it is now: looked at through the materialistic eyes of conventional neuroscience, the brain is just an object in the world, like a brick or a pebble, and it has no intrinsic ownership and therefore offers no basis for the fundamental sense that I *am* this thing, even less that I am *here and now*.

It is worth bearing all this in mind as we plunge into the interior of the sensory room where our model of the world, and exquisitely structured interactions of our self and that world, are made possible.

The Listening Post

When I think of the listening head, images come to me. Of a man standing absolutely still, harvesting the sounds and signs coming from afar. Of the small head ear-cocked into the greatness of space, reaching ever-deepening distances, forever widening its awareness, its reach extended by technologies that catch voices bounced off satellites from the far ends of the earth. The collective human head, like a great radio-telescope, further out in space, further back in time, listening into what the poet Rilke called 'the endless message that arises out of silence', to the very origin of that which made humanity and its listening ears possible.

When we think of hearing, or indeed of any of our senses, we can look at mechanisms; then we can see the miracle; and finally accept the mystery. So I shall say something about the mechanisms of hearing, briefly gasp at our miraculous powers of discriminating sounds, and finally glimpse at the mystery of the transformation of vibrations in the air into a universe of sound.

Mechanisms, then – how beautiful they are. The ear has three parts. The outermost part is the pinna plus the auditory canal. The middle ear is a cavity – the tympanic cavity – filled with air. It is separated from the external ear by the eardrum (which goes under the formal name of the tympanic membrane). It houses a chain of three minute bones or auditory ossicles known as the malleus (or hammer), the incus (or anvil) and the stapes (or stirrup). (Interesting how human beings name parts of their own bodies after the tools that, because of their unique relation to their bodies, they are able to manufacture in order to magnify the powers of those same bodies.) And then there is the inner ear, which is one of the most extraordinary structures in a body made of extraordinary structures and which, at the very least, deserves a few paragraphs to itself.

The inner ear is a labyrinth within a labyrinth. The outer bony labyrinth is an exquisitely carved space in the temporal bone, just above and behind the pinna, filled with a fluid called perilymph. It has three major components: a central chamber or vestibule, three semi-circular canals, and a spiral tube which is called the cochlea because it is like a snail's shell. Suspended inside the bony labyrinth, but occupying only a small part of its space is the membranous labyrinth, which is filled with another fluid – endolymph. Each of the semicircular canals contains a semicircular duct. These open on to a sac in the vestibule – the utricle – which itself connects with another sac, the somewhat tautologously named saccule. Inside the cochlea is a coiled tapered tube, the cochlear duct.

Each of these components of the membranous labyrinth is furnished with sensitive endings that, when activated, trigger nerve impulses that travel up the eighth cranial nerve to terminate in the brain. This nerve is responsible not only for transmitting impulses triggered by sounds but also for information related to equilibrium. The utricle and saccule detect the direction of gravitational forces so that the head knows what position it is in. They are equipped with otoliths ('ear-stones') embedded on a sensitive membrane. These record the position of the head rather in the way that a drop earring may measure one's deviation from the midline. Confused signals result in a ghastly sense of dizziness or vertigo.

Acceleration in space and rotation of the body cause an inertial lag of the endolymph in the semicircular canals and this stimulates sensory hairs. Children love experimenting with this by spinning round and round. The continued movement of the endolymph after they have stopped spinning results in the persistence of the sensation of movement. The consequence is an exciting or nauseating disconnect between the visual and proprioceptive evidence of stillness and the vestibular report of continued movement.

The most amazing of the structures of the inner ear is the cochlear duct, which is responsible for hearing. The principle by which this operates is similar to that for the structures that ensure equilibrium. The relevant sensory endings are immersed in endolymph. Vibrations – originating from sounding air – transmitted through this medium stimulate the hairs in cochlear hair cells. The shear on these cells opens up ion channels which result in electrochemical events that eventually become nerve impulses passing to the auditory cortex. The cells are located on the organ of Corti, which has been described as 'the body's microphone' and lies on top of the basilar membrane, which follows the spiral of the cochlea. Everything up to this point has served only the purpose of amplifying the sound waves reaching the ear about twenty-fold.

Now the sound waves are received and the process by which they are perceived begins.

We owe much of our understanding of what happens next to a Hungarian engineer, Georg von Bekesy, who specialized in the design of telephone earphones. On the basis of some lovely experiments in reverse engineering, he came to understand how the organ of Corti distinguished different pitches, the very basis upon which we differentiate and recognize sounds. He found that the basilar membrane on which the sensitive organ of Corti is located resonates to different frequencies at different points along its length. High frequency or high pitched sounds selectively vibrate the membrane near the entrance to the snail shell. Low frequencies travel further along the helix before shaking the membrane sufficiently to excite the hair cells in the organ of Corti. This is the so-called 'place' theory of pitch perception.

Something else is needed to secure the very fine discriminations we make between pitches. Only about a dozen hair cells are associated with each distinguishable pitch and it would be difficult to imagine a mechanical resonance of the basilar membrane that had a sharp enough spatial discrimination. One mechanism that has been suggested is that the firing of hair cells either side of the minute area where resonance is at its peak is actively damped down – so-called lateral inhibition. In this way, the potentially spread-out vibrations are squashed to a sharp peak, narrowly confined to a minute area only a few hundredths of a millimetre across.

The hair cells themselves deserve an awestruck glance. There are 16,000 to 20,000 of them set out in rows on the basilar membrane which follows the spiral of the cochlea. Minute movements of the membrane trigger the hair cells to provoke action potentials in the nerves to which they are attached. Each of the hair cells has about 100 'stereocilia', leaning against each other like sticks on a bonfire. When the membrane wobbles in time to the waves of sound, the stereocilia wave around like sea weed and send impulses to the brain. The conversion of the movement of the stereocilia into nerve impulses – electromechanical transduction – involves a long chain of molecular events, which engages signals across and within cells.

The mechanisms operating through these exquisite structures, amplifying the sound waves, separating their different frequencies, and translating their mechanical movements into the common currency of the brain – nerve impulses – is necessary to secure the miracle of ordinary hearing. And it is a miracle. We identify natural sounds such as the rustling of leaves, the tread of an oncoming predator, birdsong and the thousand distinctive voices of water; we understand the bruit of artefacts – the noise of traffic, the swish of cloth, the whistling of a kettle; we hear, analyse and understand the speech of our fellows whose voices we can pick out from thousands; we enjoy song and the music of a hundred instruments. And how good we are at this! We can hear how dry those rustling leaves are; whether the water is being doled out in big drops or small; the texture of a piece of cloth that is being stroked; the sourness or warmth in a tone of voice; the plangency in the collective orchestral voice as when, in a Bruckner symphony, a private ache is unpacked to a public monument of delicious harmonies.

The composite sounds of everyday life are received in terms of their different frequencies – and duration and temporal pattern and waxing and waning or steady loudness – and then synthesized into the sound that we know. Neurophysiologists expect one day to understand how this synthesis happens in the brain. Before we express scepticism on this score, let us examine one particular feat of hearing that may stand for all the rest: the reception of speech.

An utterance is a pell-mell of phonemes. Phonemes are units of verbal sound, such as 'b', 'i' and 'g' in 'big'. These can be recognized for what they are only if they are located under general categories, based on a guess as to what the speaker is saying. The same sound may, in different circumstances, realize different phonemes; and the same phoneme may be realized in different sounds. The identification of phonemes – across all sorts of variations between speakers and within one speaker at different times – is pattern recognition that relies on different cues. In order to know whether what we are hearing is, say, a vowel, we have to hear that it is a genuine tone made up of rhythmic sounds; to hear that it is an unvoiced consonant we have to perceive that it is non-rhythmic. Between these two classes are sounds that have an underlying rhythmicity even though they are accompanied by a non-rhythmic friction noise. At different speeds of speech, volumes and tones of voice, all the diagnostic cues are reset. This may be very radical, as when you talk and sob at the same time or speak with a hot potato in your mouth. And yet I can still understand you.

The acoustic processing necessary to extract the sequence of phonemes is only the first step. Stress and tone have to be detected – so that, for example, I can tell whether you are making a statement or asking a question, whether you are pleased or annoyed. What is more, the phonemes have to be appropriately clumped together to form full word sounds. And then there is the small question of holding together the entire statement, silently echoing in my head, so that I can see what you mean. The rate at which these operations take place – and they require mobilization of a vast knowledge of language – is dizzying, as will be evidenced from the fact that a word such as 'big', that hardly outlasts a blink, has three separate phonemes.

Let me pursue the miracle of hearing a bit further with another example. When I hear a sound, I immediately know the direction it is coming from. This depends on the different timing of its arrival at the two ears. A sound coming from the left will reach my left ear before it reaches the right. Localization will be based on comparative timing. The difference, however, even in the case of a sound due left or due right, will be a matter of perhaps as little as a millisecond. And yet I have no difficulty tracking the movement of an object – as it passes from left to right – by the relocation of its sound. This is particularly impressive, given that (as we noticed in chapter eight) sounds, unlike sight, do not create a space of their own that locates its inhabitants: their loci are constructed out of the experiences of other senses.

Here is another miracle: our ability to pick sounds out of a background of other sounds. This is the so-called 'cocktail party problem' – in which I have to listen to a particular conversation when there are other equally noisy conversations demanding my attention, plus music that fills in the spaces between the meshwork of voices. This is the equivalent of picking out all the ripples associated with one raindrop from a puddle in a downpour. And then there is our ability to hear a chord, hear that it fits together as a satisfying whole and, at the same time, hear the individual notes that make up that satisfying whole. The more we know about hearing, the less we may take for granted our ability to overhear the background music that annoys us when we are trying to conduct a conversation at a noisy party.

What we know about the mechanism falls far short of explaining the miracle. And at the heart of the miracle is a mystery. We do not know, ultimately, how, in the head, vibrations in the air become sounds – informative or meaningless, background or foreground, rasping or beautiful. For, in the absence of human consciousness, vibrations in the air are just that; and audible wavelengths have no more intrinsic qualities than the ones that we cannot hear, such as ultrasound vibrations, which lie beyond the range of our ears. We do not know how physical energy becomes experienced sensation. The transformation of one form of energy to another – sound waves to the electrochemical activity that is nerve impulses – is not the same as the transformation of energy into consciousness of the energy or (as in the case of vision) consciousness of the surfaces that are its source.

Nor is it at all clear why high frequency sounds should sound as they do, on low frequency sounds should sound as *they* do. The different locations in the basement membrane that are vibrated by the sounds do not explain these qualitative differences: the 30 millimetres between the bottom of the cochlea and the top are no explanation of the difference between the sound of a double bass and that of a piccolo.

It is still a fundamental mystery why certain nerve impulses should be associated with certain experiences and other very similar nerve impulses should be associated with dramatically different experiences, for example that impulses in the visual pathways should give experiences of light and impulses in the auditory pathway give experiences of sound. We cannot appeal to the different structures they are wired up to – eyes and ears respectively – if we believe that it is in the brain itself that energy becomes experience. Indeed, this 'explanation' seems somewhat circular: you get sounds in auditory pathways and vision in visual pathways, because that is their job.

In fact, it is even less of an explanation because, as we have already noted, there is nothing intrinsically noisy about vibrations in the air or intrinsically bright about light. There are no subjective experiences built into, or guaranteed to come with, the energy that impinges on the sensory endings that make the head the marvellous sensory room that it is. Nor, finally, is there any reason why sounds – or sights – should be referred to places and objects beyond the sensory pathways, indeed beyond the body, where they are said to be experienced. Why I should see the light that comes from that object *as* that object or the appearance of that object.

The mechanisms, the miracle and the mystery are all largely hidden from us and have required extraordinary ingenuity to uncover. And it is ironical that the most prominent parts of the auditory tract are least important. We may be confident that Van Gogh's hearing was unimpaired by his famous self-mutilation. For an unimportant structure, it is impressively littered with names.[2] Adam's successors have paid loving attention to the various parts of this wonderfully rumpled structure, whose most beautiful feature is the folded rim. There is a helix, an anti-helix, a fossa of a helix or scapha, a concha, a tragus, an antitragus, a lobule (where earrings are hung) and an external auditory meatus. Of all the names, the most evocative is 'Darwin's tubercle' which designates a little swelling on the rim, and hence links the tip of my ear with the man who did more than anyone to explain how my ears (and the rest of my body) came about. Our worlds are stitched together in so many unexpected ways.

My relationship to my ears is on the whole rather low-key. The shadow cast by my pinna on my mastoid process is a prime example of those areas of my head that I do not supervise. Had I been a rugby player or a boxer I might have had more dialogue with them. It is many years since I was pulled by the ear. The most memorable instance of pinna-traction occurred courtesy of an art teacher who was also an international rugby player – front row forward. He was exasperated at my inability to draw. I was under the impression that the ears were located two thirds up my head rather than, as they are, halfway down. Mr A. made his point by pulling on my ear.

The cauliflowering of the ear may have more esoteric causes than contact sports:

In the late 1800s, when opium smoking was still popular, the presence of cauliflower... was considered almost [diagnostic] for opium use. They were the result of lying for long periods on opium beds with hard wooden pillows.[3]

There is something rather outlying, even clip-on, about ears. Those iconic ears made with engineered tissues grown on the back of the experimental mouse only emphasize how ears are a kind of after-thought. This is particularly true of bat ears, which are charmingly transilluminated when the light falls in a certain way. I have recently missed Andrew Marr's ears transilluminated by the street lamp outside 10 Downing Street as he takes us through the latest scandal, now that he has retired as the BBC's Chief Political Editor. But the most poignant example of bat ears is that of Franz Kafka. His were exaggerated by his gaunt face and thin body and he was very conscious of them: 'The auricle of my ear felt fresh, rough, cool, and succulent as a leaf, to the touch.'[4] I wonder how, if his ears had not undermined his claim to be taken seriously by passing strangers, and if he had had better luck with women, the history of modern literature might have been different.

The philosopher Ludwig Wittgenstein once observed that one could terrify with one's eyes but not with one's ears. There are times, however, when we catch sight of ourselves talking too much to someone who seems to be listening too hard. The listener becomes an auditor and we seem to be falling into their judgement as down a well. Silence may be as powerful as the loudest scream.

Taste and Smell

Compared with vision, taste is primitive. It does not of itself persuade us to posit extra-corporeal objects: we taste tastes. The tongue we taste with is, of course, solid and palpates the object that yields the taste and it is this that testifies to its independent existence. Taste is not just a matter of taste but of survival. 'Taste drives appetite and protects us from poisons.'[5] The elements in the ground floor of taste – those *Ur*-tastes which are certainly beyond dispute – are four plus one: sweet, sour, bitter, salty – and *umami*, a meaty, savoury taste due to certain amino acids or to sodium monoglutamate, which may also stimulate the *umami* receptors.

The first four make 'suck it and see' less hazardous. We are advised at once whether to spit out or swallow:

> We like the taste of sugar because we have an absolute require-
> ment for carbohydrates…We get cravings for salt because we
> must have sodium choride … in our diet. Bitter and sour cause
> aversive, avoidance reactions because most poisons are bitter …
> and off food goes sour.[6]

Even *umami*, known to the Japanese for a long time but relatively new to the West, gets the gustatory thumbs up for sound biological reasons: it drives our appetite for amino acids, which are the building blocks of protein. (Bacon speaks loudly to our umami receptors because it is a rich source of amino acids…No wonder public health warnings are impotent against its blandishments.)

The primary triggers for taste stimulate the taste buds. These are groups of 30 to 100 cells – called 'neuroepithelial' because they belong to an intermediate zone between the skin and the nervous system – located on the taste papillae. Put out your tongue and you will see little red dots, particularly at the front of the tongue. These are the 'fungiform' papillae (they are like button mushrooms). There are other, less prominent papillae: foliate, circumvallate and non-gustatory filiform or thread-like. While most of the taste buds are on the tongue, they are also found elsewhere in the mouth. You have just under 5,000 taste buds enabling you to see what it is you are sucking.

Different fundamental tastes tickle up their cognate taste buds in different ways, though the principle is the same. The relevant sub-stance, dissolved in saliva, enters the taste bud cell through a pore in its centre. If the molecular key fits the lock, it binds to the membrane of the receptor cell. This primes a cascade of events that directly or indirectly enable an influx of calcium ions into the receptor cell. Release of a neurotransmitter results. It excites nervous pathways that end up in two places: the somatosensory cortex of the brain for the conscious perception of tastes and the hypothalamus and other areas said to be responsible for behavioural responses – such as spitting out the food or calling for the chef's head.

Taste, however, is not simply a series of gustatory chords, descending fourths of salt and sweet, of bitter and sour. It is mainly smell, as anyone knows who has had a cold. Without smell, we could not taste the difference between grated apple and grated onion or a chunk of apple and a chunk of turnip. In fact, tasting differences goes far beyond what can be differentiated by taste. Flavour, which is what we are really concerned about, is the resultant of taste, smell, texture (in which the inside of the mouth reveals itself to be a exquisitely sensitive organ of touch, registering the fizzy and the slobbery and the crunchy and fudgy) and other physical characteristics, such as temperature. Uncrumbling apple crumble, cold tea, crunchy potatoes offer their tastes to us in vain. Watermelon makes up for what it lacks in taste by its snow-like texture. And perhaps sound has a role to play as well: the gravel path crunch of celery, the squeak of fried cheese.

The sophistication of our sense of taste goes beyond the intra-oral teamwork of a multiplicity of senses working together. Our tastes are a mark of our taste – in food, in wine, in friends, in sexual partners, in hobbies. Close relationships are based upon a sharing of tastes that give us an appetite for each other. We fancy each other because we fancy similar things outside of each other: our taste receptors have similar configurations. And we work hard to acquire tastes – training our palates, strenuously learning new enjoyments.

Taste very quickly adapts to a stimulus: perception of an applied substance fades within seconds. This micro-tragedy is replicated on a macroscopic scale over our lives: taste (and smell) jade and fade in old age:

> Let me disclose the gifts reserved for age
>   To set a crown upon your lifetime's effort.
>   First, the cold friction of expiring sense
> Without enchantment, offering no promise
>   But bitter tastelessness of shadow fruit
>   As body and soul begin to fall asunder.[7]

The world darkens, falls silent, scentless and insipid and finally intangible. Our only consolation is that we will not perceive it: tastelessness will not be 'bitter'.

In the meantime, we may console ourself with the profound mystery we visited when we wondered about the transformation of vibrations in the air into noise and music. Sweet and some bitter taste stimuli activate a chemical messenger called 'gustducin'.[8] This prompts an 'electrochemical dialogue' between the receptor cells which transmit messages to cells at the bottom of the bud. The basal cells can also talk back to receptor cells and among themselves. Out of this dialogue, 'once everybody has their stories straight', the data are then relayed to the brain. What is most intriguing is that gustducin is a close chemical cousin of transducin, a substance intimately involved in translating light energy into vision. The puzzle we addressed with hearing comes back to haunt us: why should the different forms of physical (or chemical) energy be associated with different kinds of sensation, particularly given that the qualities we sense are not inherent in those forms of energy, or indeed in the material world at all?

Asking this question makes synaesthesia, the slippage between one sense and another, so that we see sounds and hear colours, experience the hue of cries and the dissonance of hues, easy to understand. Cases of synaesthesia are not unusual. Richard Cytowic has recounted the case of 'The Man Who Tasted Shapes' an individual for whom different shapes were associated with different tastes.[9] There has recently been described the case of a woman who had vision-touch synaesthesia: when she saw someone being touched, she experienced the same touch on herself.[10] This synaesthesia was importantly restricted: she experienced it only in relation to the touching of human beings. Of course, it is only humans, or in relation to humans, that we are touched by what we see. (The other's gaze is like a sight returned as touch, as we have seen.)[11]

★

All our senses are mysterious but smell the most explicitly so. There is something profoundly paradoxical about a sense that is at once primitive and yet so personal. Of all the senses, it is the most enchanted curator of the past. Certainly, when I am packing my bag for the next world, I will find space for a few scents: the odour of rain on hot pavements, the pine scent that exhales happiness, and one or two other fragrances that are connected with the more predictable of my appetites. Smells are so intensely evocative precisely because they are everywhere and nowhere, because unlike sights and sounds and touches, they connote rather than denote, because they say nothing out loud but offer ambience and murmur, because they are untethered to any object and hence are even more remote from our intellect with its Propositional Awareness. Because they are things we notice without noticing that we notice, they capture a whole world – the world to which they, and once ourselves, belong.[12] Worlds, and the emotions that transform them, are a bit like smells: they steal up on us and steal away.

The most famous smell in literature was the one that contributed 90 per cent of the taste of 'the little piece of madeleine' which, dipped in tea, transported Marcel Proust from the staleness of weary middle age to a lost paradise of his childhood, when the world was new and names had their full freight of meaning, leading him to reflect that:

> when from the long-distant past nothing subsists, after the
> people are dead, after the things are broken and scattered,
> taste and smell alone, more fragile but more enduring, more
> insubstantial, more persistent, more faithful, remain poised a
> long time, like souls, remembering, waiting, hoping, amid all
> the ruins of the rest; and bear unflinchingly, in the tiny and
> almost impalpable drop of their essence, the vast structure of
> recollection.[13]

We perhaps underrate smells because we think they belong to a world we have left, an animal world which is a nexus of scents, a network of olfactory gradients, whose inhabitants spend much of their lives following their noses. We have stood up and raised ourselves above the pongy, fragrant earth and now exchange loving or lustful glances rather than sniff each other's bottoms. Nevertheless, smell plays a greater part in our lives than we realize. At the very least, it informs us of the quality of the air we are breathing (in particular whether it is polluted with smoke), and of the presence of other living things whom we might wish to eat, to avoid or to recognize. But we can smell other things, for example, the emotions of our fellows such as fear, contentment and their state of sexual arousal.[14] Women are better at this than men. They can discriminate more reliably between armpit swabs taken from people watching 'happy' and 'sad' films. Emotional literacy, it seems, is mediated by the nose.

Smell depends on the olfactory membrane at the back of the nasal cavity. The membrane is about the size of a postage stamp and contains 10 million receptors. (Dogs, for whom smell is much more than recreational, have a billion or more receptors.) These are nerve cells with hairlike cilia that trap passing particles for smell analysis. Odorant molecules – which have to be small enough to be volatile – key into different receptors. Exactly what forms the 'wards in the lock' is unclear. There have been heated exchanges as to whether molecular shape, the ability to diffuse across the cell membrane, the triggering of small local electrical currents by deformation of the membrane, or molecular vibrations are involved.

Work that has attracted universal acclaim (and the Nobel Prize for Physiology or Medicine in 2004) was that of Richard Axel and Linda Buck, who made major discoveries about the nature of odorant receptors and the way the smell brain is organized.[15] There are about 1,000 different types of receptor cells which can respond to more than one smell, making us able to recognize over 10,000 distinct odours, scents, fragrances, pongs and stinks. Sniffing the air, olfactory peering, is richly – sometimes too richly – rewarded. Axel and Buck found that each olfactory receptor cell possesses only one type of odorant receptor but that these are switched on by genes. There may be as many as 1,000 different genes, accounting for 2 per cent of the human genome. The neurones from cells responding to particular types of odorant project to the same part of the 'olfactory bulb' which is the major way-station in the brain for transmitting neural activity associated with smell. Onward transmission enables activity to be combined, so that complex fragrances – for example, the freshness of early morning – can be concocted.

Often we communicate by smell without knowing it. As we all know, animals – for example horses and dogs – can smell our fear. This is sometimes regarded as a provocation, so that being afraid of being mauled or trampled brings about the very thing our fear fears. More disturbing is that our attractiveness, if we have it, is rooted in pheromones rather than our witty analysis of the failings of Mr Blair's government, our record of kind acts, or even our big brown eyes. On that score, we may perhaps relax: a liberal dousing of pheromone shaving lotion seems unlikely to redeem the bore; after all, our propositional awareness deals in the ideas of world rather than being shaped by tropism.

One of the most hotly contested areas of smell-signalling between people concerns its role in the phenomenon by which women living in close proximity are said to develop synchronous menstrual periods. Martha McClintock and her colleagues found that armpit swabs taken from donor women at a certain phase in their menstrual cycle and wiped on the upper lip of a recipient could advance or retard menstruation in the recipients depending on the phase of the donor.[16] This fascinating observation has been questioned by Beverley Strassmann,[17] who impressively under-mines McClintock's statistics; but the attention it has commanded shows how hungry some people are to believe that we are in the grip of forces of which we know nothing. There are many thinkers out there who would like to believe that we attract our mates through the pheromones our bodies emit rather than the sonnets over which we have slaved; that our lovers love us for reasons that are known only to our selfish genes and not for the things we believe ourselves to be.

The Head Senses Itself

The head sees, hears, tastes, and sometimes smells itself. Even more intimately, it feels itself directly. The breeze parting to pass around my head, picking up turbulence from my ears, may create a tactile chiaroscuro, with leeward and windward cheeks differentially awoken to themselves. This sensation, and the sound associated with it, must be one of the oldest head sounds in the world, whispering to our hominid ancestors as they waited for their prey to come into view. Cephalic self-awareness may amount almost to a self-portrait – though a somewhat broken one and executed in atypical materials. And we may raise our head's self-awareness, by biting the inside of our cheeks or playing with our hair.

The head may talk to itself. There are hummings and pulsings and ringings. They become prominent when what is out there is less clamant and the waters settle. There are after-images in the eyes, echoes in the ears, after-burning of sun-exposed skin, or the continuing pressure of a hat removed. In silence, we feel a faint fizzing in the skull, like the schhh… of a recently poured glass of lemonade. At night, after a day in the sun, I can sense a skullcap of scorch and count my pulse throbbing in my cranium. With my eyes closed, I can tell what position my head is in.

One of the cruellest of the head's private monologues is tinnitus: the skull shouting at itself. The ringing is the equivalent of a constant meaningless after-image or a permanent bitter taste in the mouth. A famously tragic example was that suffered by the great composer Smetana. The syphilis that did for him plagued him with anti-music: 'a buzzing and tingling in my ears, as if I were standing by a huge waterfall'.[18] In the final years of his productive life, he lived in a village deep in a forest whose plush of silence gave him no pleasure: 'deafness would be tolerable only if all were quiet in my head'. Nevertheless, most of his masterpieces were composed during this time, including the symphonic cycle *Ma Vlast* whose musical biography of the river Vltava has meant that this river has overflowed its banks to flood remote places of the world. Smetana even wrote a string quartet, *From My Life*, whose final movement represented his tinnitus as an intrusively dissonant note. Could there be a more heroic example of how we appropriate our bodies and suborn even the insults they throw at us into material to serve our higher purposes? The tinnitus that curdled the silence to which

Smetana's deafness condemned him is still resounding through concert halls filled with appreciative listeners over a century and a half later.

Two years after the premiere of *Ma Vlast* gave Smetana long delayed recognition, he died, insane, in a Prague lunatic asylum of neurosyphilis.

Third Explicitly Philosophical Digression:
Having and Using My Head

We are told so often when young to 'use our heads'. That commonplace instruction acknowledges two further ways in which we are related to our heads: instrumental and possessive.

Possessive first: *my* head – I own it, it seems. Ownership of my head and its parts is generally presupposed but there are circumstances under which it is made explicit: 'My head is a funny shape'; 'I had to lower my head in order to avoid banging it'; 'I've just bitten my tongue'; 'My face looks red'; 'My eyes are sharp'. 'Don't touch my head'.

Next, instrumental. Here I want to focus on those situations where I use the *material* of my head; where I take advantage of its material properties. There are some obvious examples: footballers heading a ball; street fighters nutting their opponents – a particularly literal form of tête à tête, to which we shall return; the deliberate use of one's head to block somebody's view. Let me alight on just one example, perhaps the earliest instance of being encouraged to use our head in a game we have all played: Peep-bo or, more properly, Bo-peep.

The rules are simple, though the metaphysics runs deep. Bo-peep requires two players: one who is to provide a surprise; and the other, an infant, who is to be surprised. Mother withdraws her head and then suddenly exhibits it, either at an unexpected place or at an unexpected time, or both (this is the 'peep' bit) and, at the same time underlines her returned presence, represented by her head, by crying out (the 'bo' bit). Notwithstanding the simplicity of the game, infant pleasure may be sustained over many repetitions, as weary and (let us be honest) bored parents know to their cost. The source of the pleasure, however, is easy to understand. Mother and infant are playing with the presence and absence of significant others. The pleasure comes from relief at the return of the absent one, at the resolution of uncertainty as to where and when the return will take place, and from a small, manageable fright, provided by the interjection. (If the dose of the latter is miscalculated, there may be problems. Like all games, it can end in tears.) What is remarkable is that the mother is utilizing the visibility of her head as the instrument by which the game is played. That we can deliberately exploit this physical property says a lot about the Byzantine nature of our relationship to our heads.[1]

Ownership of my body and utilizing parts of it are closely connected. The many-layered modes of 'having' encompass awareness of our carnal being as a tool-kit. As noted earlier, this probably owes its ultimate origin to the sense of the hand as a tool.[2] The 'utility' of the hand spreads into other parts of the body, themselves used as tools. But the tool-hand not only instrumentalizes the rest of the body (which both directly and indirectly through artefacts reinforces the instrumentality of the hand) but also awakens conscious ownership. In the overwhelming majority of cases, it is only a *part* of the body that is a tool – a tool that is used by the rest of the body, or is experienced by the body diffusely as an agent. The body, in short, differentiates into tool and tool-user.

It is important not to exaggerate the explicitness of this divide.

We don't use our body when we are hurriedly and efficiently tying our shoelaces in the way that we might use a device for helping us to tie our shoelaces. In fact, the instrumental attitude to our own bodies is more likely to be evident when a new, or for some other reason difficult, task is being undertaken: when the task requires particular efforts of either precision or power – as when I am learning the art of heading a ball. Then my head becomes a tool in proportion as I will it to perform in a certain way; in proportion as it resists utilization; in proportion as it is deficient. The 'toolness' of my head may also be especially apparent when it undertakes a task usually assigned to another part, when, for example, I use my mouth not to eat but to break a piece of thread or to get a cap off a bottle. This is true of other parts of the body; as when I use my shoulder to batter down a door or my toes to hold a fishing net I am trying to repair.

We may use other parts of the body to modify the head-tool, cupping our hands behind our ears to enhance hearing or shading our eyes from dazzle or pinching our noses to seal ourselves off from a pong. This active manipulation of our sense organs is a crucial aspect of the instrumentalization of the body: it is a vital step on the road from the unfocused, unsystematic gawping of beasts to the deliberate, systematic inquiries undertaken by humans.

Manipulating our bodies enhances our awareness of them as *objects*. This is particularly true when we seem to be struggling against their limitations as in effortful activity; but also on more low-key occasions. When, for example, I put on a pullover, inserting my head through what is called (by an interesting transferred epithet), the 'neck'. Or when we purposely make ourselves visible (if we think we look good) or invisible (if we feel we look stupid or are unwanted). This highlights another mode of possession of our bodies: through our awareness of our appearance to others.

'My appearance – What a possessive!' as Paul Valéry might have said.[3] It is something I may wish to flaunt, to conceal or to pretend to be modest about. I can manipulate it in all sorts of ways. Assumed expressions or postures, make-up and jewellery, expensive clothes with designer labels – all of these reflect the notion of my body's visibility as an asset or a liability. Some parts of the body – of women's bodies in particular – are particularly likely to be seen as 'assets'. In extreme cases, bodily presence shades into a self-presentation that is more or less pure asset management. Concern about 'my' appearance is not of course confined to 'looks'. We are equally concerned about how our body may smell, or, when we speak, sound, or even, when we shake hands, feel.

The feeling that I look silly or in some other way deficient may cause me to blush. Blushes – whose profound origins we have already explored – may in turn become a secondary asset or liability that needs to be managed. They may or may not be charming depending upon how close they are to the idea of female vulnerability and how far they are from looking like a physiological reaction and/or a rash. The complexity of blushing – such that in certain cultures a permanent state of blushing may be simulated through a 'blusher' powder – is a striking indication of the many layers of our possessive relationships to our heads, evident in our layered apprehension of that head's own appearance.

Consciously being in relation to its actual or imagined appearance is, rather as with suffering, a way of *being* our body, in which possession and being possessed keep on changing places. *What* it is that is appearing, how I appear, is at least in part in the keeping of certain others, or the General Other. That is why even this blush, which is warm on my face, betrays me, and gives me over, to others. The dark, general, organic interior of my body that is relevant to physical suffering is replaced by the dark, general, social exterior in the case of 'my' appearance and the social suffering it may cause. The fact that our appearance is in the (uncontrollable) judgement of others can become an obsession.

My appearance is thus at once closest to me – inseparable from my surface (try lifting a blush off a cheek) – and beyond my control. While my head's intelligible appearance to others is necessary to enable it to express my thoughts and feelings, it also places it at permanent risk of emitting inadvertent signals and being misunderstood. My appearance is therefore always on the cusp between using and being used, between having and being had. A possession, in short, by which we are at risk of being possessed.

There is a further sense – the obverse or dark underside of the aspects we have been exploring – in which my head may be thought of, and even experienced as, a possession: I have (or some cultures believe I have) a right to 'dispose of' my body as I wish. I alone – or my next of kin, assumed to be acting as my agent by proxy – may donate to the nation the cornea through which I am presently looking at these words. In some cultures, I may dispose of my body-life through seeking euthanasia when the suffering associated with an untreatable fatal illness becomes more than I wish to bear. Killing ourselves is the ultimate expression of the notion that we, insofar as we are identified with our body, are also not identified with our body; that our body – and the life it sustains – is our possession to do with what we wish.

This is the supreme expression of the paradox of 'having' a body that is an instrument of one's will, since it is not, in the last analysis, separate from the operations, or even the agenda, of that will. We are able to will the end of our willing. As the twentieth-century Catholic Existentialist philosopher Gabriel Marcel says:

> 'Having' as such seems to have a tendency to destroy and lose itself in the very thing it began by possessing, but which now absorbs the master who thought he could control it.[4]

And he adds: 'I can only *have*, in the strict sense of the word, something whose existence is, up to a certain point, independent of me.'[5]

When we consider the body as a whole – and it is born and dies as a whole – the possessive relationship fades as possessor and possession merge. While I may possess parts and attributes of my body, I am not able to possess it as a whole; this is a level at which my body and I converge and my limits increasingly become identical with its boundaries. This is echoed in Sartre's assertion that '[the body] is the master instrument for all other instruments. But it is the instrument we cannot use because we *are* it.'[6] This is not accurate because, as we have already discussed, there is a hierarchy in the body, such that one part of the body may be used as an instrument by another part, or act against the diffuse background agency of the rest.

My body may possess me in overwhelming illness. Multilayered, multistranded 'having' comes to an end as it is reabsorbed into bodily being, rather as a multitude of rivers drowns in the sea. Having and using are reabsorbed into the inutile corpse. Nietzsche's observation that 'we are possessed by our possessions' applies absolutely to the primordial possession of the body. To modify Shakespeare: 'Farewell, thou are too near for my possessing.'

chapter ten
Head Traffic:
Eating, Vomiting and Smoking

Breadhead

For those who like to claim (as I did when I was fifteen) that we are 'just animals', eating (and its obverse, defecation) seems to confirm this – at least to casual observation. Sit yourself comfortably in a café and watch a crowd of strangers filling their faces. How utterly, comically primitive it seems: into a hole in the head is stuffed the organic means to organic life. In due course, stuff will be expelled from the other end, to make way for the next lot of stuff. How very like a worm. The fact that, as you look round, you see toddlers and elders, the untutored and their teachers, revolving their mouths in harmony, licking their lips like dogs relishing Chappie, using the tongue to move food around the oral cavity and to dislodge it from places where it has lodged, picking their teeth, wiping their hands, seems only to underline the basic nature of the action. No training, knowledge, expertise, is required to eat: it comes with the starter pack.

Civilization, it seems, has not civilized eating; or not, at least, changed its essential nature. Richard Dawkins, who made gene-eyed evolutionary theory his key to understanding humans, and informed us that we were in the grip of selfish genes, admits that 'to force a naive Darwinian interpretation on everything we do in our everyday lives would be an error'. We are, after all, 'totally surrounded by artefacts of our own civilization. The environment in which we now live has especially little to do with that in which we were naturally selected.'[1] He adds that, however, 'We can still make a simple Darwinian interpretation of things like hunger and sex drives, but for most of our questions we have to employ re-writing rules.'

This seems to be true when we take a distant view and reflect on the astonishing metamorphoses of the means by which organisms acquire the nutrition required to maintain themselves and their order in the teeth of a universe tending towards the complete disorder of thermodynamic equilibrium. The life of plants is, as W. H. Auden observed, 'one continuous, solitary meal'.[2] Somewhat more sophisticated are those organisms that filter water or the air for specks of nutrition. Yet more so the earthworm, that drags fragments of plants into its burrow and sucks food into its mouth by means of its pharynx. Next there are the herbivores, the gatherers, that wander in search of food which they alight upon or reach out for: eating is guided by perception. Then there are the carnivores, the predators whose food has ideas of its own – most centrally that of not becoming food – so that they have to deploy strength, speed, and several modes of know-how (including knowing how to cooperate with one another) in order to fill their stomachs.

After such metamorphoses – from absorbing, aspirating and filtering, to reaching out, biting and chewing, and thence to hunting and gathering – the further transformations of eating by humans seem minor. But they are not. The more carefully we look, the wider yawns the gulf between the way humans and all other animals interact with food. You can see this as you observe what is going on in a café.

The food, first of all, is cooked (how well or badly is irrelevant) or, more broadly, processed. The culinary revolution preceded the agricultural revolution: humans cooked food before they grew, or domesticated, it. The structural anthropologist Claude Lévi-Strauss made the contrast between the raw and the cooked a fundamental marker of humanity, a symbol of the contrast humans felt and asserted and elaborated, between culture and nature. This symbolization was further elaborated within meals that had complex, grammatical structures, reflecting principles that dictated which dishes could go with which and the order in which dishes would be eaten. The sequence of soup, meat and two veg, and jam roll and custard is a perfectly formed English culinary sentence. By contrast, the family dog, tied up outside the café, will eat anything in any order: his meals are a stochastic series of mouthfuls. And then there are the technologies – the recipes, implements, packaging – that lie behind the simplest of meals.

Secondly, the meals are attached to times of the day. This café is busy because it is 'lunchtime'. For animals, eating, if it is broken up at all (and is not a continuous activity, or passivity), takes place when an appetite encounters something it wants to eat. Human eating is regulated by public, shared time, not solely by private hunger. Eating in turn contributes centrally to structuring the day. The scarcely formatted time of toddlerhood is structured as much by 'breakfast', 'lunch', 'tea' etc as by the succession of night and day. In many cultures, the year is importantly mapped out by feast days. The timing of meals, what is more, has been a marker of class differences. In the past, the upper classes typically breakfasted late (about 10 a.m.), as befitted their leisured status, thus distinguishing themselves from the lower orders who ate very early before going off to work.[3] In France, the day traditionally revolves around the hinge of the enormous midday meal and the post-prandial siesta. Even the speed of eating is a cultural issue. The drive to ever faster food has provoked a 'slow food' movement, dedicated to reaffirming the intrinsic social and, indeed, spiritual value of eating, of eating with others, of sharing savoured experiences, and symbolically applying the brakes to a world in the grip of a delirium of joyless consumption.

And yes, of course, meals are, thirdly, social occasions. The socializing may be more important than the food. The café is full of people who have met to eat together; or, rather, are using eating together as a peg on which to hang a meeting, a pretext for other purposes: gossiping, plotting, courtship, developing a friendship, passing time, and so on. Darwinians would like to minimize this extraordinary elaboration of the social dimension of eating. A chimpanzee reaches out for a banana and offers it to another chimpanzee. That is feeding behaviour. I invite you to have a meal on me because I like you. You, knowing that I have just taken on a hefty mortgage and wanting to be liked by me, falsely declare that you are full after the main course, and decline the offer of a pudding. That, too, is feeding behaviour. But the differences between the two modes of feeding behaviour are more important than the similarities which the use of the same phrase tries to conceal.

The shared meal is a node in an almost boundless nexus, which reminds us of, fourthly, the miles that are gathered up around the table. For animals, food miles end when the food is encountered; for humans they are just beginning. The miles between the first hand that touched the food and the head that ingests it may run into thousands. The items that end up in the stomachs being filled in the café will have been sent from China, Brazil, Stoke-on-Trent, South Africa – many times the distance between the hand and the mouth. The implements with which they are eaten will also have arrived on the table after a long journey from a country where goods are manufactured cheaply because of lower wages. And, of course, those who are gathered around the table will have come to this spot – on foot, by car, by plane – by virtue of many thousands of small actions necessary to carry out that expression of sustained purpose called a journey. As the meal is concluded and the bill is requested, we are reminded of more lengths of food miles: those we have covered in order to earn the cash to pay for it. As Marx pointed, out, we are different from other animals in that we produce the means of our own subsistence and the objects that satisfy our needs are transformed into commodities.[4] All of those journeys to and from, and within, work, all of those journeys in our training for work – these are wrapped up in the money that passes out of our hands in payment of the bill.

At this point, the differences between animal and human feeding do indeed seem almost coterminous with the differences between nature and culture. The social occasion is highly ritualized in so many ways. It is no surprise that the ultimate ritual for humans hungry to make contact with the hidden world and the hidden beings they believe govern their lives consists of eating one another or their gods. The Eucharist is a fascinating atavism and invites laughter: 'They do a bloody good Body and Blood of Christ at St Stephen's' etc. The rituals are constrained not only by the succession of dishes but by table manners that are at once (largely) arbitrary and non-negotiable.

Table manners begin with the table itself, which has to be properly cleaned and robed. And then with the laying out of implements which mediate between the hand and the mouth, making the stuffing of food into the mouth seem less primitive. The ability to cope with place settings and to know which fork to use when is still in some circles one of the most powerful markers of social differentiation. Elbows on the table, dunking the wrong solid in the wrong liquid, talking with one's mouth full, also betray poor upbringing or, even worse, poor breeding, a deficiency that combines genetic with educational failure. Some infringements may be rather sophisticated: for example, saying 'Bon appetit' before a meal in a private home in France;[5] or prodding the air with a fork upon which a piece of meat has been skewered in order to put a philosophical assertion in italics, in the hope that it will carry more conviction.

Tables, table cloths (best linen, perhaps, a wedding present or holiday souvenir), place settings (that test the social climber), place mats (that portray country houses), napkins (folded in a prescribed way) underline how the centrepiece of the most ordinary dining room is a dense network of signs – biographical, social, geographical, historical – and each sign is the beginning of another endless chain of signs. Think of the way the preference for the term 'napkin' over 'serviette' or of the manner in which a tea cup is to be held (little finger offset from the porcelain) becomes a marker of a certain social class; secondarily, a means by which that social class is got hold of and lampooned; and thirdly, a quaint reminder of the era in which a certain sort of marker of social class seemed to matter and was much talked of. What huge distances we humans have inserted between the hand and the mouth and how hard evolutionary psychologists and others who would wish to biologize humanity try to conceal them!

Nevertheless, eating cannot entirely shake off its primitive origins, notwithstanding that the food is being inserted into a lip-sticked sphincter (brand carefully chosen and not tested on animals), just below the reading glasses (designer frames) of a woman who is struggling to understand a rather difficult sonnet (which may well be the prepared text in the examination that lies in her path to a teaching career), and has just had her roots dyed in accordance with the latest fashionable rebellion against convention. That child inserting a sausage into his mouth, an act of reverse defecation, bears witness to Samuel Beckett's assertion that the mouth is the anus of the face. I am reminded of how, in his film

The Phantom of Liberty, Louis Buñuel reversed the attitudes to eating and defecation: people sat around a table on chamberpots, contentedly chatting away and, from time to time, excused themselves to go to the toilet in order to eat. We cannot help being sickened when someone chokes on a breadcrumb and brings up a bolus of masticated food. And the spectacle of a fellow diner talking and eating with his mouth open, so that we are given a bird's eye view of his food-lined oral cavity, does not offend us solely because potatoes and prose make such ill-assorted mouth-fellows – and prosing is very much a secondary function compared with chewing – but because it reminds us what eating actually is.

Actually, the anti-social act of speaking comprehensibly with one's mouth full deserves some praise. After all, food dramatically changes the acoustics of the oral cavity, creating the need to regulate breathing so that breath is available for articulate expiration, while ensuring that dinner does not end up in the lungs, and requires the simultaneous engagement of oro-facial muscles in both speech and mastication. This should result in an impenetrable dialect but, for the most part, it does not. We may not enjoy conversations with individuals who communicate the busy interior of their mouths as well as their thoughts, but we can understand them.

In the end, however, opening our mouths to put in food and drink seems a comically unworthy use of heads that in so many other respects have 'moved on'. There is something retrograde about using this wonderful structure for purposes shared with chimps and sparrows and earthworms. This is, however, tricky territory. A decade or so ago, the late Robert Nozick, a philosopher of some renown, published a book celebrating the ordinary things of life. He dwelt on the mystery of everyday bodily functions, among them eating. He expressed astonishment at the fact that we open a hole in our head and stuff material into it. The unsympathetic reviewer was unimpressed. 'Has he told the Royal Society about this?' she asked sarcastically. 'If not, why not?' She had a point. Sometimes unscientific philosophizing, without the protecting veil of technicalities, can look like 'Whizdom'.[6]

Vomiting: Theory and Practice[7]

One sure-fire cure for 'whizdom' – or indeed for any whimsical or philosophical stance on one's own body – is vomiting. There can be few experiences so all-consuming. No orgasm, symphony or wartime briefing can command such total attention. Your body has you in its entire grip. There is a sense of helplessness as you are frogmarched through a series of stages beginning with diffuse sickliness; through nausea that grows in intensity with sweating, pallor and the copious salivation as the harbingers of the gathering storm; past the preliminary heaves and retches; and then on to the gross and utterly engrossing climax. There may be a minute or two for you to prepare, once the ordeal seems inevitable: time to take off some clothes, to kneel down by the toilet bowl, and to take a few deep breathes, and thus stockpile some oxygen like a hoarder faced with a coming shortage, for who knows when you will next be able to breathe freely without the danger of inhaling vomit. There is a kind of terror in vomiting: it is a shouted reminder that we are embodied in an organism that has its own agenda, and that agenda might not include breathing for the present. The unpleasantness is compounded when the stuff runs down the nasopharynx and the delicate membranes are scorched by half-digested nutriment stewed in hydrochloric acid. Waiting for the fit to pass is an archetypal expression of the waiting that is inherent in all illness.[8]

The mechanism of this almost laughably crude activity is not at all simple. Nausea is associated with decreased gastric motility, increased tone in the small intestine and a few waves of reversed peristalsis in the upper part of the small intestine: a few spots of rain before the storm bursts. Then there are 'dry heaves': spasmodic inspiratory movements against a closed airway. During this time, the furthest part of the stomach contracts a little and the nearer parts relax – as if marshalling the content to the places from which it can most conveniently be hurled into the outside world to the disgust, astonishment and concern of one's fellow humans. And then the phoney war comes to an end and the real business of throwing up begins.

Here's what you would have to do if vomiting did not happen involuntarily. Take a deep breath, close your glottis (the opening to the airway) and open your upper oesophageal sphincter (the opening to the gullet). Elevate your soft palate to try to stop the vomitus from flooding up the back of your nose. Pull your respiratory diaphragm down sharply to create a negative pressure in the chest, which will cause the oesophagus and the sphincter between it and the stomach to open widely. At the same time, contract the muscles of the abdominal wall, squeezing the contents of the stomach which will shoot up the gullet via the mouth to the alarmed, fascinated and nauseated outside world.

Vomiting is choreographed from the brain. In the brain-stem – the little stalk that links the cerebral hemispheres with the spinal cord – there are two vomiting centres, one neural and one hormonal. The vomition centres receive complaints from the gastro-intestinal tracts and other viscera via the autonomic nerves that report something amiss – such as over-distension of the stomach, which, as we know from daily experience, is a very potent stimulus for vomiting. They also respond to certain psychological stimuli such as fear, offensive odours and nauseating spectacles, to disturbances from the balance pathways (as in sea-sickness), and to damage to the brain. The chemoreceptor trigger zone by contrast responds to emetic drugs and a variety of metabolic disturbances.

The causes of vomiting are myriad. Bad or too much food

and/or drink, tablets that disagree with you, and disease of the digestive tract, are the obvious causes. But heart attacks, kidney failure, liver disease, and a variety of endocrine and metabolic problems may also cause vomiting. The voluntary vomiting of the anorexic and bulimic has attracted an enormous coverage in the medical, sociological, psychological, pop-sociological, pop-psychological and lifestyle press.

Psychogenic vomiting may baffle doctors and result in a large number of inconclusive tests on this organ and that. People can sometimes make other people sick and unresolved conflicts with parents or spouses may result in a career of chronic vomiting – more commonly in women than men.[9] 'You make me sick!' may be more than a metaphor, though of course vomiting is a prolific source of metaphors for all kinds and conditions of disgust. Calling for the sick bag when we are forced to listen to a load of sentimental or insincere twaddle is a form of behaviour that locates us a long way from the brain-stem vomition centres and the chemoreceptor trigger zones in the floor of the fourth ventricle. Humans can make food for thought even out of vomit.

Vomiting is an ideal topic for laughter, too. For a start it is something we all fear. (There was hardly a night that passed when I was a small child that I did not worry that I might vomit and choke to death in my sleep.) What's more, there could be no more complete divorce between how things are and how they ought be than seeing a person reduced to a body in the grip of emetic convulsions and seeing his head turned into a gargoyle emitting a stinking effluvium of half-digested food. Some of the humour of vomiting is captured in the multitude of terms it has attracted: chunder, barf, etc. According to www.vomit.com there are just under 400 synonyms for the event. One of the world's favourite vomiting jokes concerns Sir Les Patterson, Australian Cultural Attaché to the Court of St James (aka Barry Humphries). Sir Les, having over-indulged on duty-free booze on a stormy crossing of the English Channel, is copiously sick over a lady holding a little dog in her lap. As he apologetically cleans her up, he comes upon the drowned lapdog and holding it in the air says 'Good grief, I don't remember eating that'.

Smoking

We do many strange things with our heads but smoking must be one of the strangest. Now that, in the UK at least, it is disappearing from public view, its strangeness is becoming more apparent. American comedian Bob Newhart beautifully captured its oddness in the sketch where he looks at smoking through the eyes of a businessman to whom Sir Walter Raleigh is trying to sell the idea of cigarettes. The businessman cannot believe his ears when Raleigh explains what smokers do with tobacco. He wonders whether the man has lost his mind. At any rate, he warns Raleigh, he is 'gonna have rather a tough time selling people on sticking burning leaves in their mouths'.[10]

How wrong could you be? Currently, there are 1.1 billion smokers worldwide, 80,000 to 100,000 new smokers a day and 4 million smoking-related deaths annually. By 2025, it is estimated that in China alone there will be 2 million deaths per year from tobacco.[11] So what is the attraction (in a third of cases fatal) of this peculiar habit?

Smoking is a way of signalling one's arrival in the adult world. The giving and receiving of cigarettes is what the sociologists would call 'a trivial coordinator' of human co-presence. Pausing for a fag – at work, on a walk, between bouts of coitus – is a way of making the pause more than the pause. It relieves 'stress'. It allows little journeys around the moment. As Calvinistic Scottish epidemiologist Thomas McKeown once said, smoking, like many of our vices, begins as a pleasure we do not need and ends as a necessity that gives us no pleasure. Nicotine is addictive, a fact denied on oath, until recently, by senior executives of the giant tobacco corporations.

It is not even, at first, a pleasure: it is an act of will that requires the debutant(e) to overcome quite a few unpleasant sensations in order to reach the fag-end of an early proof of adulthood. But it is not long before refraining from smoking becomes an act of will and the weedless hours are an ordeal of restlessness. Before long, too, some of the tissues of which the smoker is made – lungs, heart, mouth – may be smoke-tweaked into biographies of their own that are at odds with the life the smoker wishes to lead. In short, all that choosing – nipping out for a puff, liberating the packet from the cellophane with its built-in faultline, accepting with grateful thanks, lighting up, quick-last-dragging before tossing away the butt – hasten the end of choosing, bringing on an end we did not choose.

The nauseating headachy smell of public spaces broadcasts that cigarettes, pace Richard Klein (author of *Cigarettes are Sublime*), are not sublime. The metallic taste transmitted to the somewhat more private space of the mouth of the smoker's carnal partner re-inforces it. The saucer filled with fagends, ash and spent matches conveys the same message to the eye, as do the butts that leak tobacco as they disintegrate in the urinal. The ashtray and the sputum pot are obvious cousins.

To press this point a little further: the steepness of the line cor-relating the health consequences of smoking with social class[12] emphasizes the 'fagness' of cigarettes; that they are not small, chic, exceptionally neat cigars but 'ciggies' cosmetized out of the truth of their destructiveness. The demographic facts of smoking are now more truly represented by the Woodbine concealed in the curled nicotine-stained hand of the furtive subaltern than by the scented Sobranie held aloft at the distal end of a mono-grammed holder by a dressing-gowned patrician.

At the bottom of the joy of smoking is the mystery of smoke. To think about smoke is to attempt to grasp it with something no less smoke-like; to try to palpate and pluck fog with hands made of mist. (We may think of thought as the human body's subtlest smoke – or its subtlest breathing – a thought that seems to have particular aptness on a cold day which turns chance remarks to rags of steam.) Thought's adjectives are odourless and its verbs do not reflect smoke's self-transformation from morning blue to fagged-out grey, or the beautiful way smoke seems to shrug its shoulders at the passing of time as it unfolds upwards. Even so, mind-smoking is still to be preferred to the real thing, as it is butt-less, duty-free, and does not contribute to a public stink in public places or make a private one in otherwise kissable mouths. Nor does it mess with death, that limit to the thinkable.

More satisfactory than mind-smoking, perhaps, is observing smoke captured on celluloid, preferably on black and white film. The arc-light transilluminating the studio gives smoke sharper edges and firmer planes than it enjoys in the wider world, and thus transformed it plants its image in undarkened silver on the film. Such undissolving images outlast the dissolving smoke. And, come to think of it, the smoker. The cigarette scissor-gripped between the fingers of long-dead Marlene Dietrich is again and again unpacked to smoke as her life breath draws the cigarette's death-breath into her body (long turned to ash or worse). The intertwining pillars of blue-grey still climb together from the ashtray where the two lovers, also intertwined, have parked their cigarettes (in separate grooves on the rim), the silence of the smoke's ascent uttering the erotic silence of their carnal conversation – a sublime distance from the wet butt in the lavatory.

The smoker's elegance has several distinct components. There are, for example, the innumerable ways of playing the beautiful little cameo role of 'One who lights up'. The elegant avoid (for example) the use of many perfectly handy surfaces for striking matches: the brick on the pub hearth, the heel of a muck-crusted boot, the stubbled chin (though there may be a kind of grace even here). Virtuosi performances may include a pause to complete a sentence, during which the soft-vertexed triangle of flame grows down the stalk towards the digits holding it; for this injects the risk necessary for grace to be shown. However it is kindled, and whatever risks are taken, the flame must be allowed only to kiss its object and it must do so, ideally, under the lightest of inspirations from the smoker's exquisitely perfumed evening body. The most elegant modes of lighting up are those which invoke the assistance of another and allow the gift of fire to be received with grace. Ms Dietrich was once approached by thirteen men jostling in the ante-room of her attention, proffering their lighters to her unlit cigarette. Thirteen, she declared, was unlucky. A fourteenth stepped forward – Mr Ernest Hemingway – and his was the lucky strike.

There are, of course, many other opportunities for elegance. That first deep inhalation where the satisfaction of a 'craving' may demonstrate just how deep one's feelings go. The exhalation (plumed through the mouth, or emitted through the nose and forked, like the devil's beard, because noses have two nostrils) while one utters aphorisms whose tokens are cured in smoke. The dislodgement of the ash by the lightest tap of a painted nail attached to a manicured hand. The extinction of the butt made unexpectedly exquisite by preoccupation with some higher thing.

Human beings have many ways of tending their own bodies and also of tending their appearances. Smoking combines both in a very complicated way. Elegant smoking, posturing with cigarettes, is – from the first clumsy green-faced experiments behind the bike shed to that perfect evening-gloved hour in the cocktail bar of the five-star hotel – about owning, manipulating, tending and exhibiting the external surface of one's own body. Does this not say something about our relationship to the very organism that cigarettes corrupt?

There is an enduring fascination in observing the transformation of the cigarette as its glowing tip eats backwards, translating the passage of time into movement in space, with the smoke symbolizing the presence of the past and the unburnt remainder the presence of the future. And, most beguiling of all, the way smoked smoke insinuates itself between sentences and gestures, and inscribes on the air a kind of space (and time) unknown to physics. The laws of physics, in accordance with which a cigarette unpacks itself to smoke that wanders over a room, leaving the start of a stain around a chandelier and ash that crumbles over papers, carpets and furniture, are time-reversible. The laws of motion, for example, do not forbid smoke from gathering itself from the ends of the earth (even from the lungs of the blackbird in the garden who inhaled it as he paused for breath between songs) and returning to a forwardly unblazing cigarillo. But, of course, such reordering cannot happen: the Second Law of Thermodynamics stating that disorder tends to increase forbids it. This is not, however, quite the special mode of space-time I am referring to. I am thinking more of the spaces in which smoke meets words and its particles interpose themselves between the successive sounds of a sentence as they ride on the exhaled air shaped by the smoker's mouth according to the meanings she wishes to convey. And I am thinking of (or mind-smoking) the fact of human agency which makes 'smoke' a verb – and a transitive one at that so that the smoker smokes the cigarettes while the smoking fire smokes only itself – and upgrades smoke from a mere effluent of natural processes to a desired commodity manufactured to meet a manufactured desire.

Which brings us back to something that I have touched on when we considered eating. Smoking exemplifies the circuitous relationships humans may establish to their own bodies. A smoker, like a chimp eating a banana, is altering her own body in response to an appetite declaring itself in that body. But the smoker has not simply reached out for the cognate object of her need: she has bought matches and has deliberately moved to a smoking compartment. She has manipulated the laws of the material world that have the chimp in their grip.

All of this exhibits the peculiar nature of human beings: how they are points of origin in the universe, and so *do* things rather than merely suffer them; how they insert tables into time and format those tenses that physics says do not exist; how they come to themselves and even the visible surfaces of their own bodies from such distant places. In so thinking, I have drifted up to that place where plumes of smoke rub the surfaces of artefacts gathered in rooms. Here the thicket of thought grows so dense that no further movement is possible.

chapter eleven

Head on Head: Notes on Kissing

Dear as remember'd kisses after death.

Tennyson, 'Tears, Idle Tears'

A kiss – so little and so much. A kiss is about carnality, about human freedom, about our tracks through the world, and about memory and the endurance of human relationships.

It is early in the seventies of the last century. A man is travelling in a train to meet a woman from whom he has been separated by a very long fortnight as a 104-hour-a-week junior doctor. He is in love. She is in love. When they meet, they will kiss each other. All pretty straightforward. But nothing could be less straightforward than a kiss, those two heads closing in on each other – hers at walking speed, as she makes her way to the station, and his at a speed varying between 0 and 80 miles an hour, sitting in the carriage of a train.

They are impatient for this collision. His impatience has grown as the appointed time has got nearer, in accordance with the principle that time expands in proportion as we resent the distance it inserts between us and our goals. This distance, if it cannot be slept away, has to be lived through a succession of experiences and thoughts and events and actions. Strange things happen to time when distorted in the field of impatience. Zeno kicks in: the interval is halved and halved and yet there seem to be just as many moments to be got through. The fortnight, halved, leaves a week which seems as long as the fortnight; ditto for several days and a whole week; for one day and three days; and for this journey and one day. The upshot is that the two-hour journey separating one city from another seems as long as the thirteen days separating their previous farewell from the greeting to come. There is an explanation for this doleful mystery: the acuity with which we resolve time into its component tasks increases as the interval to be crossed decreases.

In a justly famous paper, Benoit Mandelbrot asked the seemingly dull question, 'How Long Is the Coast of Britain?'[1] The length of a coastline depends on the acuity of your ruler. If your estimate of length is based on a satellite picture, it will be less than that of an observer following out all its coves and beaches, whose estimate will in turn be less than that of a third observer following a snail leaving a trail of silver on every pebble. We could imagine a fourth observer using a powerful microscope to track all the ups and downs on the surface of each grain of sand. His coast of England will be longer still. From this one may conclude either that the coast has no intrinsic length or that its intrinsic length is infinite.

This is for mathematicians and philosophers to argue over. For the man in love, the important point is that the experience of waiting to meet his beloved will be protracted as the acuity with which events are observed, anticipated and recalled, is racked up. The week breaks up into separate days; the days break up into clinics and ward rounds; a ward round separates into individual patients; the treatment of a patient resolves into a large number of actions, such as history taking, ordering X-Rays, giving injections, even explaining to the patient what is going on. Each of those actions has many components. Ordering an X-Ray breaks down into filling in a form, walking to the phone and ringing for the porter to take the form down to the X-Ray receptionist. Further divisions are evident: three attempts to phone through to the X-Ray department; twenty five separate steps to the phone; eight movements involved in each attempt to get past the engaged tone; and so on.

But now at last he is underway, cancelling the space between her head and his. He tries not to think of the many components of this journey, so that this inland trajectory does not balloon like the coast of the country he is passing through. As his head is hurled through the intervening shires, however, and his face, trailed over hedges and fields, develops like a photograph as dusk falls outside, he cannot help thinking how big the world is, and how small by comparison are the heads that are to meet. The counties extruded from between their mouths are approximately a quarter of a million times the size of the two actors in question.[2] This is the size of the world their heads contain.

Of what concern, other than to the two heads aiming to collide, are these considerations of time and space? It is of the utmost concern and to us all. Because whatever else the kiss is, it is sexual, and this encourages some to think of the culmination of our meeting – to which the kiss provides the entrée – as animal behaviour. Freud, for example, asserts that

> Anyone who subjects himself to a serious self- examination on the subject… will be sure to find that he regards the sexual act as basically something degrading, which defiles and pollutes not only the body.
>
> The excremental is all too intimately and inseparably bound up with the sexual; the position of the genitals –*intra urinas et faeces* – remains the decisive and unchangeable factor… The genitals themselves have not taken part in the development of the human body in the direction of beauty: they have remained

animal, and thus love, too, has remained in essence just
as animal as it ever was.[3]

And the reader may recall Richard Dawkins bracketing sex drives
with hunger as the two remaining parts of human behaviour that
are susceptible to 'a simple Darwinian interpretation'.[4]

Yet there is no simple Darwinian account of an action that has
so many intermediate steps, of a journey to a goal that takes place
through such a densely woven medium of events; that involves, for
example, removing money from one's wallet in order to deploy
one set of conventional signs (pound notes – it is 1970) for another
(a railway ticket that was designed in Doncaster, printed in
Edinburgh, purchased in Portsmouth, stored in the pocket of
clothes appropriate to the anticipated occasion until it is yielded up
in London, and made of card created out of trees chopped down in
Sweden and dyed in Bollington, upon which is printed information
in an alphabet perfected 2,500 years ago, and regulated by laws and
bye-laws which have been fashioned as a result of the operation of
collective experience on the intuitions collected in common law).

Nor is the joy it brings simple. The journey is a pilgrimage of
one world to the heart of another. Why otherwise would he have
felt that the world around him had an extra-special connectedness,
if it had not been gathered up in imaginary consciousness of her,
awaiting him at the station? Why otherwise would the way the
coffee mounted the sides of the polystyrene container have given
him such pleasure as he recalled that this was the inertial influence
of the distant stars making life on earth the distinctive hard work
that it is? Why otherwise did the thought that the mouth into
which the sandwiches were being fed was an extraordinary multi-
tasker seem profound?

This journey, then, of the one head towards another is not propelled simply by the operation of instincts (even less mechanical

causes). The two could not meet, except by deliberate action, an explicit sense of purpose, transilluminating the material world, and the million and one steps dictated by practical reason, through which this goal has to be achieved. We are talking about knowing creatures here, operating more or less freely. Yes, freedom *here*, where animal appetite is considered to reign. Yes, practical reason, accountability, unclouded consciousness, where irrational passion is supposed to be the driver. With the biological story marginalized – though of course what is being done has biological roots – we are at liberty to examine what is going on: to look at this strange activity that is kissing (not to speak of the stranger activity to which each expects it will lead).

And so to the journey's end and picking that one face out of so many faces, that one head out of so many heads, in the busy concourse of the station. The time this takes says something about the contingency of the relationship between the object of one's desire and the desires that one has; between the one that is imagined and the one that exists out there. Thirty-six years on from that kiss, so many days, nights, so many thousands of conversations, contingency has been somewhat mitigated by a life together, by their role in making each other what they have become, in making their children and in making a shared past that is the past they each contain within them singly. Such are the ways of *post hoc* necessity, of becoming, and so being, that which one has experienced.

And so to the kiss. Kissing is both a biological and a social scandal. Biologically, it is a waste of time. The ten-second jump and shriek of our nearest animal kin, is more to the point. The genotype never managed to replicate itself through oro-oral intercourse. And socially it is deeply problematic. There are, of course, grades of kissing: mouth to epithelium (hand, top of head, cheek); closed-mouthed lip to lip; open-mouthed lip to lip; and the full French, where tongue licks tongue and ravenous oral orifice feeds hungrily on ravenous oral orifice. The passage is from the abstract and symbolic of the kiss of prelate's ring, dry as written sentences, to the total carnality of one person drowning in the absolute 'Yes' of another's open mouth.

Locating a kiss at the right place in the spectrum can cause trouble. As I write, a vicar and school governor who kissed a prize-winning girl on the cheek when he handed her prize at the prize-giving ceremony has been referred to three separate disciplinary bodies for investigation, evaluation and appraisal of his possibly inappropriate behaviour. The matter has evoked national comment.

The kiss for which our lovers were transporting their heads so many miles is driven by mutual physical attraction – as the phrase goes – but it reaches into psyches and symbols that are unknown to biology. Of all the sexual moments, it is least at risk of becoming impersonal. Which is why, so we are told, prostitutes do not go in much for kissing: they want to keep themselves at a distance from the bodies that they are pleasurelessly using to pleasure – that and the fact that it is less unpleasant sucking someone's impersonal cock. Cocks and their emissions may taste foul but at least it is a body not a person that the foulness comes from. For the pleasure of lip on lip, tongue on tongue, is not guaranteed. 'Every kiss is a conquest of repulsion' Sartre said with his characteristic tendency to put a positive gloss on things.

A full-on kiss is massively, and deliciously, transgressive, not only because it breaks the elementary rules of hygiene. Under normal circumstances, the intercourse between heads, whereby each tastes something of what is in the other's head, is quite circumspect: verbal or visual, exchanges of sentences and or of facial expressions and glances. The kiss bypasses all this. It is a regression to the carnal directness of infancy. The mouth as an organ of speech, of relatively remote communication about things brought into play through their general or abstract properties, is returned to its origin as a tactile organ.

Let us draw back for a moment and return to Sartre's kiss as 'the conquest of repulsion'. It seems as if, as Christopher Hamilton says, 'in sex we suspend or overcome our normal sense of disgust'.[5] He quotes William Ian Miller:

> [S]exual desire depends on the idea of a prohibited domain of the disgusting. A person's tongue in your mouth could be experienced as a pleasure or as a most repulsive and nauseating intrusion depending on the state of relations that exist or are being negotiated between you and the person. But someone else's tongue in your mouth can be a sign of intimacy *because* it can also be a disgusting assault.[6]

This takes us halfway to full understanding. But the kiss goes deeper than the conquest of disgust or the recruitment of the disgusting to the expression of intimacy with another person. To see what it demonstrates, we need to think about the relationship between organic being, experience, knowledge and our relations to others.

The central event in our early life is a gradual appropriation of our own body as our own, followed by the development of a consciousness that locates us in a world that transcends our physical existence and our bodily sensations. We enter a space of general possibilities, of a pooled awareness that is owned by no one in particular, a universe of facts, of knowledge which belongs to a community of minds. This is the world in which 'That it is warm' distances us from the feeling of warmth. And it is in this space, this world, where, for the most part, we meet with and interact with others: the world of Propositional Awareness and, more narrowly, of discourse – of general abstract signs, of conversation, instruction and instructions; of wall-to-wall words; of interactive headwinding and of frozen headwinds printed on paper or screens or magnetic media. When we make love, we bypass this world and derive our experiences directly from the experiencing body. I touch you not with my words or my exemplary kindness, but with my hands. I touch my lips with your lips.

This unmediated interaction reverses the journey that took us from organic infancy to the full acculturated mature human being. We shed the layers of worldness that usually enclose us when we interact. This is dramatically announced in sexual kissing, where the organs that enable us to interact through sentences are returned to sentient flesh that touch and rub one another. The place where we are at our most sophisticatedly human, our speaking mouth, our expressive head, is returned to its organic status.

This is not, of course, the whole story. We remain, as we kiss, knowing animals, aware of the place we have come from and of the distance we have travelled. We do not forget our point of origin. The would-be organic interactions are profoundly symbolic. When, to quote Paul Valéry's M. Teste, 'we play at being silly beasts together',[7] we are conscious of playing and of the permissions, the overridden prohibitions, which make this such a privileged game. Carnal knowledge is neither purely carnal, nor purely knowledge, but transgressive modes of direct sensory experience in which the mediations of knowledge are bypassed. This is why the kiss, for all that it makes direct contact with the other as organism, still retains the ghost of a dialogue that acknowledges the other for what she is as a person. That long journey to the kiss was a personal journey, undertaken by a conscious agent, a human being. And it is why those who kiss often close their eyes:

> Lovers, approaching to kiss,
> instinctively shut their eyes

before their faces
can be reduced to
anatomical data.[8]

So there we have it. A kiss, casting light on the nature of time,
on the infinite mediations of our daily life, on our strange, hybrid
nature as carnal beings who are more than our carnal being, on
transgression, on the interweaving of lives, and on joy.

chapter twelve
Headgear

There will be time, there will be time
To prepare a face to meet the faces that you meet;

T. S. Eliot, 'The Love Song of J. Alfred Prufrock'

The Burden of Appearance

One glance in the mirror and I have joined the largest and least exclusive club in the world: of those who are dissatisfied with their own appearance. We are the only animals who are conscious of, have, judge or inhabit, our own appearance. We see how we look through the eyes of another – a general other or a particular other or a general other represented by particular others.

Judgement goes beyond what I see in the mirror; beyond what is revealed by the light that bounces from my head to the mirror and from the mirror back to the eyes in my head. For my sense of my appearance, and of the appearances I must keep up, floods the interstices of my sense of my self. The light that lights up my appearance is informed by past light, itself stained by distant lights collected from near and far and arrested in words – in spoken and unspoken thoughts, in frivolous gossip and careful appraisal. I feel judgement being passed on the large spot on my nose, on my clumsy attempts at friendliness, on my performance on the ward round, on my track record as a doctor or a father, on my life. That foolish smile I felt just now on my face that betrays me as an idiot straddles the physical surface of my face and the infinitely imbricated surface of my self. A helpless prisoner of the other's gaze, I am cornered by my appearance. I give as good as I get, of course; others are skewered on their own appearance by my weary, knowing smile.

What strange creatures, then, we are to have as one of our most important and closest possessions our own appearance, so that maintaining face is one of our most constant preoccupations and losing it a catastrophe that may make life unbearable and suicide the obvious career move. Embarrassment is more often feared than death. No wonder we spend so much time on our appearance and amending it.

The natural place to start is with what a hippie informed me in the 1960s was 'man's natural plumage' – the hair.

Natural Plumage

The head secretes hair in all sorts of places. Not only the cranium and the temple but down the neck, on the chin and cheek, on the eyebrows, in the ears and up the nose. Even on the upper surface of the nose.

Minor outbreaks of hair may punch above their weight. Eyebrows, for example. Maybe it is a question of location, location – the same principle that makes a broom cupboard in Bromley a bigger deal than a mansion in Mansfield. Doubtless, being placed immediately above the eyes confers unfair advantage, so that a couple of mere caterpillars of bushiness can almost define a face. Earbrows lining the outer edge of my pinnae would command less attention. Eyebrows not only upgrade its glances, making them more threatening and authoritative, but also have wider effects; for example, creating an impression of benign unkemptness, or facial deshabille. Their arches perfect the arch look. And when they are momentarily raised to accompany a very mildly subversive suggestion – 'Fancy a swift half?' – they can convey an inexpressibly amusing tone of consciousness. Hair 'sprouting' out of the ears – it doesn't just sit there but comes at you in its profusion – is a gift to novelists. It is an economical way of conveying the yuckishness of adults to the young child. As for nostril hair… I was not aware of mine until I visited a barber in Turkey. When his scissors were inserted up my nostrils, I was doubly anxious: for the surgical integrity of my nostrils and the medical health of the previous nostrils into which the scissors had been inserted.

And so we come to larger fry: the beard and/or moustache and/or sideburns. It is extraordinary that this 'natural plumage' is central to so many conventions of appearance, so that it can stand for a human group or an historical epoch. The distinctive moustaches of 1970s porn stars, San Francisco gays, Balkan peasants, military men and Mexicans seem almost metonymic of their ways of life. A coincidence of toothbrush moustaches – which made many laugh who would otherwise be terrified – enabled Charlie Chaplin to produce powerful propaganda against one of the most hideous tyrants and ghastly regimes in the history of this animal that keeps up appearances and for whom esteem is so central. No wonder the care of this seemingly trivial item of natural plumage has requisitioned so many activities – trimming, shaping, colouring, waxing, combing, setting and styling – and tools and materials, such as dyes, bleaches, powders, waxes, ointments, combs, brushes, razors and scissors.[1]

Beards come in even greater varieties. They can serve as markers of masculine maturity, of social standing and even professional role. Amongst those professions where beards are most frequently de rigueur, the priesthood is paramount. Gurus, swamis, imams, rabbis, and bishops and lesser beings in the Orthodox Church, are all expected to sport beards of varying length, shape and hygienic standard. God's prophets are overwhelmingly bearded. This reflects the fundamental assumption that the connection between God and man is best mediated by those who sport the XY chromosome: masculine beardedness is a guarantee of the clarity of soul, a transcendent wisdom tuned to eternity, unlike the cognitively and spiritually suspect bearers of the XX chromosome.

If I am rather obviously postponing the moment when I turn my attention to cranial hair, this is because I am somewhat depleted in that respect. There are compensations. While all my days are bad hair days (if epithets can be fastened to the non-existent), I am spared the anguish of not being able to shape it according to an elusive idea of perfection, or the irritation of seeing a carefully constructed quiff deconstructed by pointless rain or a comb-over lifted at its hinge by mindless wind. Combing, parting, washing, shampooing, conditioning, curling, straightening, cutting, shaping, styling, plaiting etc take very little of my time. And so I am free to address a philosophical problem famously associated with baldness. This is the Sorites, little-by-little problem – or the problem of 'the heap'. If you have a heap of grains of wheat, taking one grain away cannot make the difference between being a heap and not being a heap; and yet if you repeatedly take one grain away, sooner or later, what was a heap is no longer entitled to be called a heap.

The Philosophy of Hair

In 1980, according to the photographic evidence, I was not bald. By 1990, I most certainly was. At some time between those two dates, I must have gone bald, as my poll thinned hair by hair. It would, however, be daft to say that (for example) I went bald at 4.15 on 16 March 1987.

The least interesting, because least illuminating, response is to say that the problem arises because 'bald' is a vague term that does not specify a definite amount of hair loss, so that the question of when I became bald boils down to an epidemiological question about the pattern of usage of certain terms. I became bald, for example, when over 50 per cent of a panel of English speakers would rate me as bald. Perhaps the problem could be solved by a committee that stipulated exactly how the term 'bald' should be used. This would generate further difficulties, even if the authority of the committee commanded universal acceptance. Supposing they laid down that a man could be declared bald when he had less than three whole hairs on his head. This would precipitate further disputes as to what should count as 'a whole hair'.

A different approach, which may reach further into the heart of the matter, is to see the problem as one of an incommensurateness between the realm of articulate discourse, in which people are, for example, classified as 'bald' or 'non-bald', and that of direct sense experience. It arises because of a mismatch between factual reality, which is chopped up into a finite number of distinct categories, and unmediated sense experience, which is not.

In the case of the Sorites paradox, the shear or mismatch is not as radical as it is in other cases because it does not arise directly out of an attempt to match articulated knowledge on to immediate experience. It is a matter of trying incorrectly to map one system of articulation on to another: a more gross system of categories (bald v. not-bald) on to one more fine-grained (hair by hair) – though once we try to overcome this by stipulating what will count as 'bald' in terms of the number of hairs, we get the feeling that the resistance of what is experienced to being pinned down in what is said is systematic. That is why philosophers like to say that there is 'no *fact* of the matter' for us to worry over. This is true. But we have to dig a little deeper to say just *why* there is 'no fact of the matter'. There is no 'fact of the matter' because no fact – an item of articulated awareness – can bridge the gap between the continuum of sense experience and the level of abstract knowledge. 'Fact' is the key word here: there is *no fact of the relationship between knowledge and sentience*. Facts lie on the knowledge side of the relationship.

I want to continue digging because there is a more fundamental point waiting to be unearthed. When people ask wherein vagueness lies, they are inclined to say that it is not in things themselves but in our experience, conceptions or descriptions of them. There is a definite number of hairs on that man's head over there; it becomes indefinite only when I perceive the hairs as 'many' or refer to them as 'many'. Thinking this way, while unexceptionable in itself, leads readily to two quite serious misunderstandings.

The first is the incorrect idea that there is something intrinsically precise about natural objects – that they are *exactly*, rather than merely *roughly* themselves. This is absurd: an unobserved head of hair is neither precise nor imprecise, neither exactly nor roughly what it is; it simply is. Just as an unobserved rock on Mars is neither exactly or roughly its own size. Nor does it have an exact composition which we can perceive only roughly. Exactness arises only in the context of measurement. The head of hair does not of its own accord take up a position in a measuring system in relation to which it has a definite number of components. The measuring system has to be imported from without, for it then to have a certain quantity in relation to it.

The second is that people may be inclined to suggest that sensations are vague, as compared with everyday knowledge, or the organized knowledge that is science. In fact, sensations are not even vague; we have to get above the level of pre-articulate sensation for the notions of vagueness or precision to get a foothold. The question arises only when we are making comparisons between modes of higher level awareness – when we are comparing perceptions, knowledge, thought, discourse with each other.

Although this may not seem apparent, what we are discussing is central to this inquiry into the head: the split between knowledge, that is broken up into facts, and experience which (despite all the attempts of philosophies besotted by computational models of consciousness to reduce it to binary digits) is not chopped up in this way, not articulated, not divided into atoms. The tortured relationship between my immediate experience of myself and the facts of my case, between my experience of my head and the facts about my head, is central to this head-driven exploration of the human condition.[2] The tension between experience and what Martin Heidegger called the 'de-experiences' of the world from which we are distanced is both the curse and the glory of human being.

Hair and Identity

Let us return to slightly less esoteric matters. From the outset, I have been concerned with the question of the extent to which we are or are not identical with our heads. The answer in relation to hair seems obvious. However much we are concerned with the impact of our hairstyle on our appearance to others, we are *not* our hair; otherwise haircuts would be painful, requiring perhaps an anaesthetic, and associated with a sense of loss. As a secretion, hair is somewhere between earwax and toenails (with which they have a close affinity). Yet it is difficult to believe that there isn't something special about locks of hair. Lovers, after all, exchange tresses rather than nail clippings, even though the latter have cohabited longer with their original owners.

A moving image from John Donne's 'The Relic' places hair, the outermost otherness of our body, next to the innermost otherness:

> A bracelet of bright hair about the bone,
> Will he not let us alone,
> And think that there a loving couple lies,
> Who thought that this device might be some way

To make their souls, at the last busy day,
Meet at this grave, and make a little stay?

There could be no more aching expression of the hollow hope of a
continuing togetherness after death, than this embrace between
hair and bone; that which is me and is not me.

A recent legal argument has cast an interesting light on this
ambivalence. In the wake of the Alder Hey affair – in which
worked-up parents discovered that their children's organs had been
retained for scientific purposes without explicit permission being
sought – a Human Tissues Bill was hastily put together making the
retention of material for research without permission punishable
with a three-year prison sentence.[3] The question then arose as to
what counted as material belonging to the human body. My hair
was clearly mine when it was attached to me but seemed less
clearly so when it had been cut; that is why we do not take home
the material harvested in our regular visits to the barber. The point
was made, however, that where the hair was rooted, it was
attached to skin and skin most certainly belonged to me. The question then arose of determining the line between what did and what
did not justify a prison sentence for a doctor. It proved insoluble
and the Bill was miraculously changed to something a little more
practical.

We may, of course, be betrayed by our hair. A few hairs left at
the scene of the crime will shop the perpetrator. Our protestations
of innocence, our clean blood and urine specimens, are of no use
to us when our hair is examined, for it gives the analyst a window
of several months into our drug-taking habits. Urine and blood
reports provide statements of the current account; hair has a
longer memory.

Unnatural Plumage
Making up

Let us begin with make-up, face-painting that is woven into the long history of attempts to make the face more beautiful, more frightening or in other respects more commanding of respect. It lies deep in the history of sexual attraction, of aggression and of the hunt for status. Let us focus on making the face more attractive.

The armamentarium is impressive: mascara, eyeliner, creams, powder, rouge, lipstick, and that's just for starters. And not surprisingly; for our complexions are arrived at by somewhat accidental means and often fail to correspond to what we feel we are. A young girl with a cardiac defect and a reverse shunt of blood through the chambers of the heart may have cheeks similar to those acquired by a sea captain after a lifetime of exposure to wind, saltwater, grog and hardship. Nevertheless, making up is a very strange business. Consider rouge or blusher: the use of minerals to mimic the most volatile of human colorations, to render the occurrent blushes of the red-cheeked animal as a standing state. Lipstick unpacks its crayon to lip hours of glamour, spread over time and place and encounter. The use of one lip to spread the lipstick over the other is an egregious and thought-provoking employment of the outermost part of the mouth. The two lips cooperate in gilding each other's appearance.

Lipstick has been used to demonstrate a level of self-awareness in chimpanzees (and, more recently, elephants). When a dab is painted on a chimpanzee's cheek and the animal is placed before a mirror, it will notice the mark, and try to wipe it off. This suggests that, unlike other primates, it has a sense of self; or at least an intuition that the face in the mirror is its own, is itself.[4] The significance of this has been grossly exaggerated, suggesting that the gap between us and them is slighter than we had thought. I note only that the chimpanzee did not buy the lipstick, agonize over the colour, wonder whether it would match her clothes or be in tune with current fashion, hope it would excite her partner or shock her parents, nor did she ring her friend or colour consultant for advice.

Masks are a radical approach to enhancement, control or regularization of one's appearance. (Role theorists see the entirety of our lives as devoted to creating and maintaining a mask, to managing the presentation of the self to other selves.) And various aids – notably spectacles – may be used to convey a certain impression to others as well as helping one to sense that impression. Monocles, wire-framed glasses and lorgnettes all speak about us on our behalf. Because it is more manageable, I would like, however, to focus on a remarkable trend of recent years: an eruption of heavy metal deposits on the faces of the population.

Earrings and necklaces are now supplemented by piercings. The nose, the eyebrows, the rims of the ears, the tongue – there is no place too tender or too unhygienic to be safe from implantation. The significance of this is unclear, though many of those who are metalled in this way, just as those who are tattooed, feel that it is a mode of self-expression, a way of asserting their individuality. This is, of course, very odd: conforming to a fashion is precisely the opposite of affirming one's uniqueness; a fashion by definition requires a large enough number of people to conform to it. Fashions are epidemics of slavish imitation. Perhaps they stand for a rebellion against only part of the world – that which is occupied by those who are established. They sometimes seem to say that I am not part of *that* world.

Our faces, like all living and non-living entities, are subject to the Second Law of Thermodynamics; indeed, they are complex objects entraining a good deal of negative entropy or order and so smile on in the face, or teeth, of their own improbability. It is not surprising, then, that make-up and jewellery prove insufficient to hide the evidence of wear and tear: more radical, structural changes are required – and where there is too much to be made up, a makeover is necessary. This does not always require the knife. There are more subtle ways of rearranging the face.

In the last few decades 'BoToxication' has reached almost epidemic proportions. It is an extraordinarily bold idea. Botulinum toxin is, after all, one of the most deadly of all poisons. It is produced by a bacterium, *Clostridium botulinum*. The toxin kills by interfering with the transmission of electrical impulses across the junction between nerves and muscles. Many years of squinting and frowning and smiling and other entirely legitimate uses of the face leave deep wrinkles between the eyebrows, across the forehead and at the corners of the eyes (those poetically named 'crow's feet'). Botulinum toxin is injected into the relevant motor points so that the muscles are temporarily weakened and then smooth out from disuse, and the unwanted lines, that betray the age of the face by summing up its activity, are reduced. This is, of course, only a reprieve or postponement. The wrinkles return and injections have to be repeated and, eventually, the face has to face up to the fact that it is no longer beautiful; and to withstand the inattention that anticipates its ultimate effacement.

Fourth Explicitly Philosophical Digression:
Caretaking My Head

Irrespective of exactly how and to what extent we *are* our heads, there is no doubt that our future is bound up with their future. We therefore usually take good(ish) care of our heads and we may do this directly or indirectly. We may look after our heads by looking after our hearts, and look after our hearts by putting the right food into our bodies via our heads, and engaging in exercise calibrated, as we noted earlier, by the outpouring of sweat that accompanies the body's endeavours to look after itself by regulating its internal environment. More directly, we may clad our heads in hard hats. More directly still, we may look after the appearance and health of our heads by grooming them.

Head grooming takes many forms: ears reamed, noses emuncted, eyes wiped of sleep, eyebrows plucked, face washed, make-up applied or repaired, hair combed. I shall focus on that most ordinary of *aubades*: washing the face. We have all been washing for so long, it is easy to overlook its ramifications.

Leaving aside soap – that 'magic stone' as Francis Ponge called it[1] – the silken networks of water in containers and pipes, and less silken network of regulations that link the personalized wash-basin or shower with the general rain that falls upon the just and unjust alike, and which enables us to adjust the temperature to match our physiological requirements, there is the peculiar nature of washing one's self. I use one part of my body (my hands) to operate on another (my head). I am both agent and patient. (This relationship gets even more complicated when I wash my hands.)[2]

The nature of washing is reflected in its grammar. In Greek, *'louomai*: I wash' is in the middle voice: somewhere between active and passive. In English, we can say either 'I am going to wash' (where it is assumed that the default object of washing is one's self) or 'I am going to wash myself', where that which is washed is an object, like one's car or the dishes. When we wash, we are both subjects and objects of our action. Those inclined to say that this is true of my cat which is presently running its tongue over its own arse are profoundly mistaken. A cat is neither a subject nor an object: it is not an embodied subject treating its own body as an object. This is because it does not assume its own body in self-consciousness as its own, or as itself, in the way that I do.[3]

When I wash myself, I most unquestionably treat my body as an object. The strategies I use are similar to those I use when cleaning other objects. The cotton buds that ream the cerumen out of my ears are also deployed to remove biscuit crumbs and other detritus between the letters on the keyboard. Rubbing my head first wet, then soapy, then clean and then dry, reminds me of its mulish objectivity, its brick-hardness, its barely veiled opaque otherness. One of my daily *memento mori*. And when I brush my teeth, the scrubbing is crudely mechanical.

This latter act, however, can be foregone only at the cost of feeling undesirable all day, making one wish to be at a distance from other heads which is incompatible with most of the interactions necessary for day's work. Let this action, sustained for the prescribed two minutes therefore, stand for all those many acts of grooming that mark the littoral zone between the private night and the public day. The reasons for brushing one's teeth are legion, but one is paramount: to stop one's teeth rotting (or looking as if they are rotting). This action – whose aim is not to bring something about but to prevent it from happening – plays straight into the notion of one's own body as an entity that needs to be cared for, that has built-in decay which has to be forestalled.

Out of that intuition, as applied to teeth, have grown entire technologies, rooted in science and applied through craft – of preventative, conservative and corrective dentistry. We import into the most personal place in the world the third-person awareness that is knowledge. What could be closer to us than our mouths, what could be more first person, given that, in kissing, even the mouth of the other is elevated from third to second person? Nevertheless, we accept that, at any time, lightning strikes of pain commanding absolute attention, may awaken in our mouths as a result of tooth decay. And we are willing to part company with our teeth when we judge them as no longer cost-effective. The coldest science, it seems, can find a habitation in the warmest places of our body.

All hail, then, to toothpaste which delivers both oral deodorant – or counter-odorant (for the price of eliminating personal smells is to be the bearer of general ones that carry chemical or brand names) – and abrasive material, grit in an excipient of jelly. Systematic scouring of a part of one's body – upper teeth front, upper teeth back, upper gums, lower teeth back, lower teeth front, lower gums, and special visits to cavities and recalcitrant bits of material – makes two minutes a long time, as almost as long as in the Two Minutes' Silence when so many dead have to be remembered as a congealed and blurred referent of a theoretical grief.

A large loop is coiled up inside the small circuit of toothbrushing. It girdles a good deal of the known world. If one is abroad, it may be necessary to make sure that one has an adaptor plug, suitable for the country being visited, so that one can (among other things) keep the electric toothbrush charged. This hi-tech mode of interaction between the brushing hand and the brushed mouth takes in many miles and a rather large number of abstract concepts, including the ten thousand that are incorporated in the rechargeable battery. The batteries, descendants of the Voltaic pile discovered in far-off Italy, are made in the EU and have warnings in three languages. And then there is the massive theoretical infra-structure of toothpaste: techniques of whitening, tartar detection, antibiosis etc. And, nearer to home, discussions as the proper mode of squeezing the tube, the replacement of the cap, responsibility for ensuring a supply, not to speak of all the ingenious technology that is necessary to make sure that the patriotic stripes are kept apart. How many heads have got together so that I can clean my teeth?

Thus grooming – that most intimate and immediate attention we pay to ourselves at our most literally embodied – entrains great conceptual and literal journeys, huge swathes of the natural and human world, quite unlike the situation when the cat's tongue licks the cat's arse. Nothing intervenes between one part of the cat's body and another part when the cat washes itself. When we groom, an entire universe of conventions, products, institutions, regulations, is imported into the act via the mediator – the paste-laden toothbrush – we use.

chapter thirteen
In the Wars

Warhead

Middle-class children for the most part are poorly adapted to dealing with the casual violence of the street. A girl is insulted and she responds with manganese-rich ocular saline. A boy comes to her rescue. He raises his fists to the nearest of his tormentors in imitation of his comic book heroes. Shortly afterwards he finds himself on the ground with a soggy nose. He too is secreting saline which, since it is occasioned by a mixture of physical and emotional pain, must have had an intermediate concentration of manganese. He had forgotten that the Queensberry Rules are respected only by people who do not take fighting seriously; those for whom it is essential to physical and social survival use whatever weapons lie to hand and to hell with the rules. He had been on the receiving end of a bestial butting of British bonces, otherwise known as a 'Glasgow kiss'.

The head as weapon lacks the versatility of the hand. It also has the disadvantage that it can inflict injury only by getting up close and personal – even more 'hands on' than the fist – and so exposes itself to the kind of injury it is trying to inflict. Nevertheless, as a means by which one head may hurt another head, it has a lot going for it, and in most situations much more than, say, sarcasm doled out in Alexandrines. The bones that protect the brain are formidably tough and the parts of the cranium with thick bone and high local curvature make for good weapons. For those who have taken little notice of their own head, these include the forehead near the hairline, the outward curved part of the parietal bone, and the occiput. The trick is to use a less sensitive area to strike one's opponent's more sensitive area. It is not a good idea to go for one's opponent's teeth because they are a weapon in themselves and the skin on the thick skull is likely to come off worse than the teeth.

The head is an impressive arsenal of physical and psychological weapons. The head butt, the most primitive form of tête à tête, is only the last resort, when all other less direct weapons have failed. The ancient myth that it was possible to kill someone with a glance may be just that – an ancient myth; but looks can paralyse and effectively disarm the other, so that the lance turns to rubber and the legs to jelly. Snarls, screams, shouts, curses, threats, serve a similar purpose. Spitting may repel and bites inflict something that goes deeper than the intense pain it may induce. It is the intimacy that rubs vinegar in the wound it inflicts, a ghastly inversion of sexual intimacy, in which, as in love-making, the physical sensations force the person to be where the body of his assailant is touching him. The bite tears at the soul as well as at the flesh. When Elias Canetti imagined a world in which all weapons were abolished and only biting was allowed, Niall Ferguson is correct in questioning whether 'there would be no genocides in such a radically disarmed world'.[1]

Using the head as a weapon exemplifies the strangeness of exploiting the material properties of one's own head. We have discussed this already in relation to the phenomenology of Peep-bo, where the players trade on one aspect of the materiality of the head: its visibility. In head-butting, biting, looking with menaces, this conscious utilization of the head as a material object is taken one stage further. The head-butt seems particularly odd, precisely because it is so primitive: deliberately using the head for purposes connected, not with its amazing ability to understand complex visual scenes (though this is required if one is going to be victorious in battle), to analyse sounds such as speech (though unpacking a blur of sound into specific threats is equally important), and to entertain abstract thoughts (though an awareness that reaches into the space of explicit possibility is essential), but with its weight (about 5 kilograms) and its tensile strength.[2]

The mixture of the simple and the complex, of the concrete and the abstract, gets even more Byzantine when we think of a street fight provoked by some ancestral wrong, energized by a long matured grudge, or fixed by prior arrangement. Think of the protagonists transporting their thinking heads (going over the *casus belli*, working out the route, planning tactics) to a place where those same heads can be used as material object to damage other heads, also reduced to material objects. Similar sentiments accompany my thoughts about the use of the head in football: heading, or head-kicking, seems a disgraceful use of a structure that contains or is said to contain the most sophisticated item in the world. The verb 'to head' – except where one is 'heading off' somewhere – stands for a grotesque scandal.

The term 'Glasgow kiss' reminds us how we grasp the great world we intuit out there by means of metonyms – so that a city of two million heads, as complex and as steeped in personal and collective histories as mine is, can be taken hold of. Such stereotypes – as, for example, in the notion of the Bronx cheer (any of a variety of raspberries) capturing the derision of a worldly-wise, street-wise New York – help us make the world our thing.

Head in the Wars

The head is a fearsome weapon but it is so because the heads it attacks are so vulnerable. It is therefore rightly fearful as well as fearsome. It has its own defences, of course. The brain is entirely enclosed in bone whose strength is remarkable. On 24 December 1997, one John Evans (a professional 'Head Balancer' – there's jobs for everyone in this world) balanced 101 bricks, weighing a total of 416 lbs, on his head for ten seconds at the BBC Television Centre. Nevertheless, while the delicate mechanisms of the ear are embedded in an osseous labyrinth and the eyes are snug in their bony sockets, their safety is only relative. Goggles, ear muffs and protectors, lip salve and helmets do not fundamentally change this.

Helmets – haberdashery that takes on the character of artillery in order to provide fortification – are the very emblems of war. Their history goes back to Ancient Greece: Homer's warriors wore helmets. Representations on even more ancient daggers, the Shaft Grave-daggers (so-called because they were buried with their patrician owners) of the Mycenaean age, suggest that helmets were worn at least 3,000 years ago. (How strange that a purely decorative image of a head-protector should chaperone its memory through so much time, ensuring its relay from head to head. Little did the craftsman know that he would be talking to me – and through me to you – so many years after his death.) Apart from a break in the eighteenth and nineteenth centuries, when the use of rifled firearms with increased power of penetration made them seem pointless, helmets remained central to personal protection in war. Prince Albert, who took a keen and probably ineffective and time-wasting interest in everything, a role model for our present Prince of Wales, designed a helmet that was described in *Punch* as 'a cross between a muff, a coal-scuttle and a slop-pail'.[3]

Helmets are emblematic, too. The difference between German and British helmets in the two world wars made it easy for writers

'sition in

of strip cartoons to indicate the affiliations of their characters, even without the aid of bubble-entrapped cries of 'Achtung!' and 'Blimey!'. The helmeted hordes, their headgear merged to a met-alled sea, stand for the mass wars of the first half of the twentieth century, 'the age of hatred'; for the militarized totalitarian state and total war. The heads under the helmets seem empty of conscience, of individual will, of anything approaching thought. We see the mineralization of the person, the individual reduced to a unit of aggression, a sorrow-dealing zombie. How attractive, then, is the image of the discarded helmet, lost in the long grass, woven with nettles, providing a home for honey bees.

No helmet can cancel our fundamental vulnerability, which war does not create, only exploits and exacerbates. We are always hurting our heads. The pain of a banged head must be one of the sensations with which humanity is most deeply familiar. As one gets older, the sting of injured scalp skin is salted with the humili-ation of a momentary reversion to childhood, as one is interrupted in mid-flow by a pain that makes one walk up and down clutching the injured part, oo-ing and ouch-ing, invoking excrement, sexual intercourse, and deities at random – aware, in the midst of all this, that one is cutting a comic spectacle. How truly did Shakespeare's Duke Senior say of elemental causes of suffering that

> 'these are counsellors
> That feelingly persuade me what I am.'[4]

This is but the outer circle of a hell whose circles are countless. Further in is the fall that renders one maculate like a character from a cartoon; the head-first flight through the windscreen that leaves one scarred for life and alters every relationship one has with others, mainly for the worse; the blow that leaves the bloody nose – especially handy for assailants – and, if heavy enough, a broken and permanently damaged one. (This last is also a metaphor for damaged pride, hopes and occasion for rueful reflection. Voters in by-elections are always giving governments bloody noses.) Ears can be torn off, or cabbaged, by repeated blows. Savage swipes across the side of the head may render one permanently deaf, with senseless ringing in the ears into the bargain. Eyes poked with fingers, darts, sticks, grapeshot may be transformed into 'sightless pits of pain' (to borrow C. S. Forrester's phrase). And we have still not reached the innermost circle.

For all its privileged protection, the brain is as vulnerable to damage from without as it is from within by strokes and infection. The natural experiments of traumatic brain injury have cast some light on the way the brain works and on the frightening vulnerability of our very selves. This dependency of our ordinary functioning on the brain has been testified to by many thousands of carefully investigated case studies. One of the most dramatic was the case of Mr Phineas Gage, possibly the nineteenth century's most famous neurological patient.

Mr Gage was a railway worker who had an unfortunate encounter with a steel rod that ran through his skull. This event, as the result of which he lost a lump of his frontal lobes, changed him from a purposeful, industrious worker, even-tempered and impeccably mannered into an evil-tempered drunken drifter. As Dr Harlow, the physician who reported his case, described him after his accident:

> Gage was at times pertinaciously obstinate, yet capricious and
> vacillating, devising many plans of future operations which are
> no sooner arranged than they are abandoned... a child in his
> intellectual capacity and manifestations and yet with the animal
> passions of a strong man.[5]

In short, he had completely lost the ability to direct or control himself.

Further support for the centrality of the brain comes from the many ordinary observations that indicate that the nick my brain is in and the nick my mind is in are closely correlated. A bang on the head, with damage to the brain, may remove vision, impair memory or alter personality. All of this suggests that vision, memory, personality – everything from the most primitive buzz of sensation to the most exquisitely constructed sense of self – all depend crucially on the functioning of the brain. For the neuromythologist this means that the mind or soul is housed in the brain; that within the brain is to be found everything that is required to put together sensations, perceptions, emotions, selves and persons. This is, of course, to confuse necessary with sufficient conditions.

Nevertheless, every knockout blow is a reminder of how our ordinary consciousness is contingent on things that lie outside of our control. In my medical career I have seen sufficient road-traffic accident victims to require no further reminders. One example lingers in my mind thirty years on. A wedding party had been involved in a head-on collision with a lorry. The groom was brought in first. He was in deep coma, which was hardly surprising: his skull X-ray showed a cranium cracked all over like the shell of a boiled egg that had been dropped. The sloppy brain within had large collections of blood on the outer surface; and there were numerous small haemorrhages where the brain stem had been sheared off from the cerebral hemispheres. The groom was still in his morning coat and his buttonhole was on the floor next to the trolley, trodden under the feet of the frantic medical staff. By the time the rest of the wedding party had arrived he was dead.

It is so easy to extinguish consciousness for ever. There is an appalling asymmetry between the processes required to develop, to nurture, and to educate a human head and those sufficient to destroy it. The bloodshed in Uganda, Cambodia, and Rwanda, using the crudest of weapons, has illustrated this truth in the most gruesome possible way. The low-tech approach underlines the horror of the act. In Uganda, Idi Amin murdered, or caused to be murdered, between 100,000 and 500,000 people (the most reliable figure is 300,000). Many of the murders were carried out by out-of-control militias whose favourite approach was to lie their victims down by the roadside – this provides a relatively firm surface – and then smash open the head with a sledgehammer. One blow and glances are put out, hearing is extinguished, smiles come to an end, and a world is popped. The killings were so numerous that a systematic approach had to be adopted to the removal of bodies from the Nile to prevent them clogging up the hydro-electric system. One evil head can so order the world that he can requisition the smashing of so many others. Against the simplest weapons, heads – especially those of children and babies –are helpless. Nothing can hide; not even the cochlea snug in its bony vault. It is impossible to think through the fact that it is so easy to kill someone.

Beheading is a little more sophisticated, requiring a considerable amount of strength, an appropriate weapon (sword or axe) and some training. Present-day judicial beheadings favour the sword. In Saudi Arabia, they are running at nearly 100 a year and are highly organized:

> The condemned of both sexes are given tranquillisers and then taken by police van to a public square or a car park after midday prayers. Their eyes are covered and then they are blindfolded...
>
> Dressed in their own clothes, barefoot, with shackled feet and hands cuffed behind their back, prisoners are led to the centre of [a sheet of blue plastic] where they are made to kneel facing

Mecca. An Interior Ministry Official reads out the prisoner's name and crime to the crowd of witnesses.

A policeman hands the sword to the executioner, who raises the gleaming scimitar and often swings it two or three times, before he approaches the prisoner from behind and jabs him in the back with the tip of the sword causing the person to raise their head. Normally it takes just one swing of the sword to sever the head, often sending it flying some two or three feet. Paramedics bring the head to a doctor, who uses a gloved hand to stop the fountain of blood spurting from the neck. The doctor sews the head back on, and the body is wrapped in the blue plastic sheet and taken away in an ambulance. The body is then buried in an unmarked grave in the prison cemetery... Saudi executioners take great pride in their work and the post tends to be handed down from one generation to the next.[6]

Tranquillizers ensure that the head, looking ahead for the last time, will anticipate its own de-bodied state, and the horror that will be required to achieve it, as through a blurred window. No such small mercy was permitted Mary, Queen of Scots. Because the muscles and vertebrae of neck are tough, it took three blows to hack through her neck in Fotheringay Castle in 1587. An assistant held her hair to prevent her from moving. The result was, as always, extremely gory, given that large arteries and veins, providing succour to the head, are severed. It is this, as much as anything else, which has caused beheading to fall out of favour, with shooting, lethal injection, the gas chamber and the electric chair, being preferred. They provide less stark reminders of the horror that is being perpetrated.

The mechanical destruction of one head by another is an absolute betrayal of the human world that heads, collectively, have created, howsoever this is laundered by legalistic language, or 'justified' by the reasons of state. The sledgehammer bursting open the skull of an assailant – a single act extinguishing an entire world – is a siding with the mindless materiality of the world from which, by our shared efforts, we have distanced ourselves.

chapter fourteen

The Dwindles

'The Dwindles, a term used by general practitioners
to denote – 'failure to thrive in older patients', with
'deterioration in the biologic, psychologic, and social
domains... weight loss or undernutrition... and lack
of any obvious explanation'.

Ran Han, Mark Benaroia and Barry Goldlist,
University of Toronto Medical Journal, 1999, vol. 6(3)

Speak now, and I will answer;
How shall I help you, say;
Ere to the wind's twelve quarters
I take my endless way.

A. E. Housman, *A Shropshire Lad*, xxxii

Your head came into being, and has continued to live, in the teeth
of its own improbability. It is an exquisitely differentiated, ordered
structure and hence alien to a universe that is tending towards thermodynamic equilibrium, towards undifferentiated uniformity. This
– the decrease of order – is the direction in which time's arrow flies
and its flight may be slowed, but not arrested, by the meticulous
self-repair of those tissues of which we are made. In the end, repair
mechanisms themselves fall into disrepair, so that time, used or
wasted, which has no use for us in particular, will waste us. We are
accidents waiting, sometimes fearfully, to unhappen.

In due course, this head of yours will not recognize itself. It will no longer look at itself with varying degrees of dissatisfaction in a variety of mineral and human mirrors. It will be as unhaunted as a boulder or a turnip. It does not know when, of course. It does not know which, if any of the futures it plans for itself, will have a future to house them. This ignorance, while not exactly blissful, is a blessing.

There are, however, plenty of warnings of what is to come; of our helplessness before the dwindling and dusking that will reverse our kindling into life and light. Increasingly, fatigue will mist over the mirror of consciousness. 'Our little life is rounded with a sleep' – assisted (to judge by the television viewing habits of many elders I know) by a soap. Our head nods, not in affirmation but in acquiescence, as we drift towards the ultimate passivity. Sitting up and taking notice – and who can forget that wonderful moment when an infant gains head control and starts to peer or crane? – gives way to sitting down and taking less notice. The head lolls, weary of its own weight, leaving its impress on the pillow.

The head is bowed, the horizons narrow, defeat looms. The senses lose discrimination and range. There is a narrowing of interests, of the scope of responsibilities, of the range of conversation. Appetite, ambition, desire, give way to apathy; inclination to disinclination. The number of friends, colleagues, enemies, diminishes; the sense of 'losing it' – of losing reach, grasp, grip – becomes ever more present. The world starts to close in. The last holiday abroad, the last walk in the countryside, the last foray to town, the last stroll in the park, the last visit to the post box, the last climb of the stairs. Life-space shrinks to a few rooms, one room, to a bed, and to the body itself: the bubble of the world, inflated since we awoke as infants from our bodies, slowly deflates. The head's world – whose boundless scope dwarfed the head – shrinks to the size of head out of which it grew; the toes are undiscovered; and there remain only sensations haunted by the increasingly fragmented voice. The Big Sleep draws the cerements over our life.

But let us first of all look at the Little Sleep.

Sleepy Head

> I have come to the borders of sleep
> The unfathomable deep
> Forest where all must lose
> Their way, however straight
> Or winding, soon or late;
> They cannot choose.[1]

You are too tired to think and yet cannot stop thinking. You can't sleep, distracted as you are by the disturbing spectacle of the man opposite you in the train, helpless in his stupor. His head lolls, he is oblivious of the world, saliva trickles from the corner of his mouth.

If you were told about a lifelong condition in which, for several hours every day, you lost the ability to perform voluntary actions and were subject to compelling hallucinations, you could be forgiven, even if you were an atheist, for praying to God never to fall victim to such an illness. In fact, this affliction is universal. Indeed, it is those who do *not* suffer from it who are to be pitied. The condition in question is, of course, the regular, mandatory, interruption of wakefulness by sleep; set-piece sleep, that goes beyond the minute doses of oblivion and self-fade afforded by daydreaming and those thimblefuls of darkness sipped in blinks. Our days begin and end with sleep and the arc of our life rises out of sleepy infancy and dips into sleepy old age.

Sleep has many names, reflecting different facets of this profoundly mysterious state: 'shut-eye'; 'putting out zeds'; 'kipping'.[2]

Shut-eye

The thought of open-eyed sleep is terrifying, invoking the lidless gaze of snakes, the waking dreams of the mad, or the counterfeit sleep of the spy; the sleep, in short, of the unsleeping. Shutting the eyes is a necessary condition of the repose of creatures such as ourselves in whom sight is so dominant. Given that over nine tenths of the sensory information we receive comes from our eyes, and given also that sleep is an extreme state of being uninformed, it is hardly surprising that, in preparation for sleep, we create a private night within the day, a personal darkness within the public darkness of the night.[3] The sleepyhead, dreaming of who knows what, is a striking reminder of the opacity of our heads.

The lowering of lids may be involuntary, as when sleep seeks us out rather than vice versa. The effort of keeping them open may give the impression that the eyelids, weighing at the most a few grams, are 'heavy'![4] The image of propping them up with matchsticks makes intuitive sense, though only in jokes is the notion taken literally. The link between sleep and closed lids extends even to flowers. Daisies are 'Day's eyes', and when their petals are folded, they are fancied asleep. When night is come, when Earth lies 'all Danae to the stars', then 'sleeps the crimson petal'. The sleep of flowers is nonsense, of course, since plants have no wakefulness to interrupt; but it is such beautiful nonsense that one is reluctant to sacrifice the sleep of these unconscious beauties on the altar of literal-minded truth. Indeed, the poet Rilke chose these lines for his epitaph:

> Rose, oh pure contradiction, joy
> Of being no one's sleep under so many lids.[5]

Putting out zeds

'Putting out zeds' is an altogether lighter matter. So complete is the abdication of volition associated with sleep that sleepers may lose control even of their own airways. The oropharyngeal pathways, slackened by deep inattention and a radical failure of the will, may drop into the pharynx of the mouth-breather, and the soft palate and vocal cord vibrate with each phase of each breath. The result is another headwind, this one the plainest of plain chants. It has all the monotony of breaking waves and none of their liquid beauty.

The metonym of representing sleeping through one of its manifestations – snoring – is further elaborated by representing snoring through just one of its sounds – 'z'. Perhaps 'z' was chosen because of the 'z' in 'snooze', but it is actually rarely heard. None the less there is a custom, honoured in cartoons and children's comics, of depicting sleepers as tethered to bubbles containing several instances of the 26th letter of the alphabet. Perhaps it is appropriate that the end of coherent sense should be marked by the end of the alphabet.

Kipping

The culture of sleep is many-layered. It is typically assigned a time of day, a room of its own, accessories (bed, curtained windows, shaded or extinguished lights), and a distinctive mode of dress. Any one of these items could have been turned into a verbal handle for sleep. The bed, however, has priority: it is 'time for bed' when we are sleepy rather than 'time for nightdress'.

Nowhere is the connection between sleep, sex and bed drawn more cunningly than in the verb to 'kip' and in the usage 'getting some kip' for sleeping. A 'kip' was originally a hut or a 'low' alehouse; and then a house of ill-fame, or a brothel.[6] Its meaning was extended to signify a common lodging house; subsequently narrowed to refer to a bed in such a house; and then widened again to mean a bed in general. This has ended with 'kipping' signifying 'going to bed' and (since sleep is the prime reason for going to bed) 'going for a kip' becoming slang for having a sleep.

*

Sleep itself has many varieties: broken, unbroken; restless, peaceful; dreaming, dreamless; deep, shallow; long, short. The morning's account of how we slept sometimes seems like a report on the inner weather. We speak of having slept well or badly – almost as if sleeping were something we have *done* rather than a state of inaction. We sometimes complain (or boast) of merely having 'snoozed', 'dozed' or 'napped': numerous are the allotropes of unsatisfactory sleep. The appropriately blurred semantic fields of the words map the soft-edged zones partitioning the various states of somnolence. They remind us that sleep has more than one dimension. It has depth (or depths) as well as length: snoozing is shorter-lived; dozing is shallow; and napping both short-lived and shallow. What sleep does not have is width; this is reserved for wakefulness. We announce that we are 'wide' awake – thus capturing in a single word the dilated pupil, the unshuttered lens and open orbicular sphincter, and the 360-degree solid angle consciousness of the fully alert citizen able to hold down a job and entitled to vote. But we do not talk of being narrowly asleep, or of narrowing towards sleep as twilight encroaches on the field of awareness and the snow of inattention falls ever more thickly.

We must not lose sight of the unnerving nature of sleep; and of ourselves who submit to it or seek it out. There is no gap of wakefulness, however brief, that does not take us to another shore; or to another, dislocated, place in the fractal coastline of ourselves. We never wake in quite the same place as the one in which we fell asleep. The person who resolved, when setting the alarm, to get up after a half-hour 'power nap' has remarkably little authority over the same person who, viscous with fatigue and thirsting for more sleep, gropes for the snooze button.

Awakening undusks, reverses the blurring, the fading, the directionless toppling, the global undawning. There remains the fear that we may not come round; that in our sleep we may come to harm – be assaulted from within or without our body – and our return to the world of the waking will be barred. To succumb voluntarily to sleep is to test to the limit our trust in our body, in the person with whom we share our bed, in the car in which we fall asleep driven by another, in the world.

And there is a more intimate trust: a trust in one's own consciousness. We have faith not merely that we shall in due course return to wakefulness but that, in the interim, our consciousness will not transform itself into something unbearable, a pathological awareness from which we may mine centuries. Sleep may bring dreams and in dreams the lightest thistledown of fact may unpack to a forest of triffids. The slightest hint from one's own body may build a prison whose very walls ooze terror, or shame or simple silliness.

Terror of the self-spawned spirits that haunt the dark underside of consciousness is not the only reason why 'trying to keep awake' must be one of the commonest activities of humankind.[7] Our timetabled responsibilities and our physiological need for sleep do not always match one another. To fall asleep at the wrong time puts one in danger of injury (in the jungle, in war, at the wheel) or of disgrace (at work, at a party, on a courtesy visit). We all have our favourite strategies and it is part of the massive esoteric self-knowledge we carry about with us that we know which will work best: Pro Plus tablets, fresh air, stretching to flood the brain-stem reticular formation with sensory information, sexual fantasies – the menu is endless. Sooner or later, however, we are overcome. Our head can no longer conceal its state from the heads around it.

Tiredness spreads over the waking world. Drowse mists over and foxes the mirror of consciousness with discontinuities. Eventually, an irresistible deliciousness in the arms and legs persephonically guides the head to its underworld.

If the best strategies for resisting sleep amount to an esoteric self-knowledge, so too are the ways of getting to sleep. For all of us some of the time, and for an unlucky few of us most of the time, that knowledge is insufficient. If we begin from a blur of over-whelming drowsiness, then falling asleep is as easy as falling off a log. If, on the other hand, the starting point is the harder-edged knowledge that one *ought* to sleep because it is time-tabled, then sleep will retreat as fast as one advances. We have all known the frustration when sleep plays with us: sleepiness refuses to thicken to full-blown obnubilation, while any endeavour at useful activity prompts the return of sleepiness. (How subtle the difference between sleep and sleepiness, between the full-blown state and its abstracted adverbial mode!) The connection between the ache of tiredness and the balm of sleep seems broken. Lying on our stom-achs, with the sheets reporting our weights back to us, passive, like a fallen body outlined in chalk on a pavement, we perform an inventory of the places we seem to inhabit, and the sensations that report them. And so we wait – to fall through pillows of mind-snow, longing to drift unwards.

Even when the world conforms to our desires, it does not do so instantaneously: there are many steps to the goal; each step takes time; the step from one step to another is the site of a thousand digressions; and these steps in turn are processes composed of processes that have their own tempo. We have, as the philosopher Henri Bergson said, to wait for the sugar to dissolve. It is not surprising, therefore, that there is waiting even here, in the ante-room of consciousness, as we endeavour to collude in our own dissolution.

To the would-be sleeper, the alarm clock's ticking sounds like a large- mandibled insect nibbling the leaves of time, a reminder of the final denudation, when warmth, joy, presence and being are pruned to nothing; of the dismal road to the Big Sleep. Every small problem seems urgent, huge and insoluble. The minutes pass, sieved through the capillaries of fine-grained preoccupation: 'going over the same things again and again'. You try to slip deeper into your body but are prevented from doing so by an aching gut, a crick in the neck. In vain do you petition your own consciousness to relieve you of the office of being yourself. To sleep, Novalis said, is to mate with oneself. It may be deeper than this. Perhaps becoming boundlessly self-absorbed is to undergo a private death of self- completion and world-extinction, from which one will have one's private resurrection. It is, after a fashion, a completion of meaning, a closure of the perpetual openness in sense.[8] To be unable to sleep is to be denied return to the womb of a completed self; to be refused that quiet harbour where the vocative falls silent.[9]

There is no solitude as dense as that of one who lies awake next to another who is profoundly asleep. The waking, Heraclitus said, share a common cosmos but each sleeps alone. We may sleep in the same bed but we do not sleep together, only side by side. The one who is awake, sees the other's apartness in her face closed in sleep, in her enclosedness in the dream that makes her eyes swivel under their lids, and causes her to murmur to one who is not there. Her breathing, her snoring, sounds as remote as that of the wind through a pine forest. It belongs to another world. As another gust of wild autumn wind sends the bin-lid rolling like a penny over the patio, such insomnia says: each is alone, each shall be closed off from the other, each is opaque, our shared meanings are frail, and we all die alone and when we are dead, it is forever. Sleep, in the end, comes even to the sleepless, often just before the alarm is due to ring, and wraps the insomniac in quilted oblivion that will shortly be ripped open.

> The tall forest towers;
> Its cloudy foliage lowers
> Ahead, shelf above shelf;
> Its silence I hear and obey
> That I may lose my way
> And myself. [10]

Wasteful Time

> Where wasteful time debateth with decay
> To change your day of youth to sullied night
>
> Shakespeare, *Sonnet XV*

The succession of night and day speeds towards that flickering, stroboscopic blur that the traveller in H. G. Wells' *Time Machine* experienced: the loitering days become hurrying years become fugitive quinquennia. We are propelled through an ever-increasing probability of death. Although 'being old' is in part socially and psychologically constructed, there is no part of the head untouched by the ageing process: the neurones in the cortex are fewer; the hairs sprouting out of the ears are greyer; the nose grows more bulbous, the lens of the eye more opaque; the receptors at the top of the cochlea are less responsive to the vibrations that are mysteriously translated into sounds, voices, music; and so on.

The most superficial marker of ageing is also the most symbolic: wrinkles. The cruel metonym for older people – 'wrinklies' – and the description of anti-wrinkle creams as 'anti-ageing' remedies, testifies to the centrality of skin ageing to our perception and experience of getting old. 'The cutaneous changes that occur with increasing age are universal and for the most part universally unwelcome'.[11]

> I look into my glass,
> And view my wasted skin,[12]

They are due not solely, or even primarily, to the slings of time's arrow – or to injuries. These may contribute, of course, as do the habitual postures of the face. The 'parallels' delved 'in beauty's brow' – the frontal corrugations on the forehead – may be the product, as I remarked earlier, of many surprises, or they could reflect a lifetime of supercilious attitudes. The transformation of activities into not infrequently irritated effort, with narrowed eyes as a metaphor of focused attention, leaves a deposit of crow's feet, whose claws converge on the outer canthus of the eye. A lifetime's frowning produces those vertical lines above the bridge of the nose. And then there is the bitter irony of laughter lines – the creases around the mouth – that testify to the sheer frequency with which we have noted a discrepancy between how things are and how they ought to be, and nervousness as to how they might be, that has provoked much polite and nervous aerobic and anaerobic mirth.[13] Truly has it been said that, by the age of forty, we have deserved our faces if only because we have had such a hand in making them.

The most potent factor in skin ageing, however, is exposure to the sun – the selfsame star that made life, and our life, possible and thus, remotely, fathered us. If you want to be persuaded of the importance of photo-ageing, ask a friend if she will show you her unwrinkled non-sun-exposed areas such as the buttocks. There is also the relative paucity of wrinkling in black skin, which is greatly photoprotected by melanin. Photo-ageing thins the epidermis, the uppermost layer of the skin. Most importantly, however, it results in the loss of the collagen which provides the skin with its tensile strength and structural integrity. This makes skin more susceptible to shearing forces, so that it can be stretched and not bounce back. Sagging follows and – since stretched skin has a greater surface area to be packed into the same space – to wrinkling. Brown 'age' or 'liver' spots – known as 'actinic lentigines', or more colloquially 'coffin spots', or more poetically 'autumn leaves' – are also the result of photo-ageing. None can escape, not even Cleopatra. She was with 'Phoebus' [the sun] amorous pinches black' and 'wrinkled deep in time'.[14] While age could not wither her, photo-ageing most certainly did. The 'continuous, coarse/Sand-laden wind' that is time, according to Philip Larkin, is really light-laden.[15] And to complete the irony, light itself is one of the contributory factors to age-related degeneration of the macula, the most important part of the retina. The light glancing off the world, which enables the glance to see, strikes blows that are far from glancing.

What you see in the mirror seems, in the end, to have the barest family resemblance to your younger self. You hope, perhaps, that he or she will be revived in the face of a grandchild you will one day see, eagerly facing the boundless world with an unwrinkled face.

Bonehead

> That skull had a tongue in it, and could sing once.[16]

Bone reminds of us death because more than any other tissue, it outlives our death. Our skulls are there to be picked up and exhibited by others. Bone is not, of course, dead. Indeed, it is very active in replacing itself and, in some cases, generating blood – its antithesis – out of its marrow. But it is as if the skull stands for unchanging reality beneath the pell-mell of appearance, the wooden trunk beneath the deciduous leaves. It is what remains when a voice fails, a world goes out and silence falls. It speaks from the far side of that silence: of a future when this head and this hand will be no more aware of each other than the stone hand is aware of the stone head of Rodin's *Thinker*, or one pebble senses another lying on top of it.

Not that it is entirely pleasant to think of the process by which the face and scalp are unpeeled from their bony supports and the brain and tongue and other inner parts are removed from their container. Algor mortis, rigor mortis and livor mortis are followed by decomposition. Then there is autolysis – the breaking down of the tissues by enzymes and other materials released from those same tissues. The body tenderizes itself by pre-mastication with minute chemical teeth. This is followed by putrefaction, the work of bacteria. These processes manufacture gases that speak only one word – death. And then the big battalions move in: the insects, with real mouths, as thoughtlessly bent on being, and continuing to be, what they are as the head was thoughtfully bent on the same thing until it met its defeat. Blowflies gather around the puckered omicron of the unmoving lips; beetles enter the mouth as unchallenged as hitherto the tongue, its resident sea slug, whose word there was once law; maggots cause the ears to waggle; and cheese skippers aspirate the brain, now liquefied to unthinking slop.

There is order even in this destruction of order. Hence the science of forensic entomology, which exploits the insect invasions of dead bodies to determine how much time has elapsed since the corpse was a person.[17] Forensic entomologists can either look at successive waves of insects; or use maggot age and development. The first method exploits the fact that a human body supports a very rapidly changing ecosystem as it passes from a fresh state to dry bones. Different stages of decomposition are attractive to different species of insects. 'Certain species of insects [usually blowflies and house flies] are often the first witnesses to a crime.'[18] Their response is more likely to be that of 'it's an ill wind' rather than moral outrage. Others, such as cheese skippers, prefer to wait until the corpse is no longer fresh and protein fermentation is under way. Some insects arrive only to feed on other insects at the scene. The second method uses the age of maggots which have developed from eggs laid by house flies and other insects that figured in the rapid response team. Maggots pass through many stages – first, second and instar larvae (identified according to the number of breathing holes); prepupa; pupa; and full-blown blowfly. The stage reached tells the forensic entomologist how long the body has been open for business; when the person went out and shut the door on a corpse.

Your skull, meanwhile, is as hospitable to these insects as it is to the thoughts you are presently having about them. That is what its dumb hardness, that you feel now, says: your head is not on anyone's side, least of all yours. It is as indifferent to your sorrows, your fears, your joys as it is to the song of the birds that might one day find it a ready-built shelter, as hospitable to the snake that slithers through your orbital fissure as to the light from which you have just constructed the image of your beloved. And not one of the creatures that grow, hop or gnaw their way through your rotting head will be the slightest bit curious about your thoughts, however privileged, original or salacious. For even if they continued to haunt that spot, your thoughts would offer nothing to a cheese skipper thoughtlessly intent on continuing to be what it is.

Whatsoever the legislation under which you live, however blameless or blameworthy your life, you will sooner or later suffer beheading, disarming, distrunking, debodying; and your body will be de-selved. Such thoughts about your head's thoughtless future are intended to awaken you out of usual wakefulness – which is what philosophy is or should be. The philosophical view endeavours to liberate us from our daily (usually described as 'petty' though they rarely feel like that) concerns. Imagining our empty skull, as a focus for our absence in the world, giving our future nothingness a local habitation, should open dormers in our consciousness, so that we take the long view and see how small and unimportant we are. But, of course, it does not work. The thought that we do not matter, that all things shall pass, that we cut a small figure in the order of universe, itself matters only transiently, itself passes, and cuts a small figure in the daily round of petty concerns that refuse to feel petty. Our thoughts about our minuteness are but minute parts of that which we would deem minute.

It is impossible to imagine ourselves absent from everywhere, because hitherto every exit has been an entrance elsewhere. The result, therefore, is a low-grade, unliberating fear, rather than a sense of the mystery of the contingency of our own existence. Even so, its death is the greatest idea this head can try to entertain – though it is occluded once we are busy dying. It would be nice to think that the black sun of death may make the light of ordinary hours more intense. Alas, it does not.

Nor does the Lucretian assurance – that we had nothingness before and it wasn't all that bad – bring comfort. For this time we are coming to nothingness from somethingness and it is this somethingness, this head, supported in my hands, 'the million-petalled flower of being here' that we do not wish to lose.[19] Lucretian reassurance is rather like the consolation, after some kinds of bereavement, that, after all, there was a period of one's life before one met the lost one, and one lived without any sense of privation. The comfort is even more arctic when the loss is of the whole world, the loss of one's self.

'Man', W. B. Yeats famously claimed, 'has created death'.[20] This is manifestly untrue: evolutionary theory tells us that man was created *out of* death. *H. sapiens* grew out of a rising pyramid of the unsurviving unfit. What Yeats really meant was that man, uniquely, has made death explicit. In us, transience, the condition of all things, has been transformed into an intuition, a black blush of tragedy, a mystery black-edged with terror. And so we have invented immortality, to cancel the certainty of physical death with the notion of an eternal life to which we are heading.

Headless Life

When this head contemplates its death, it exercises its unique ability to look beyond itself to a space of possibility. This ability carries a penalty: the sense of being an accident that did not necessarily have to happen. My head is simply a piece of matter in which mattering has only a temporary residence. Pain – that is at once utterly alien and deeply familiar – broadcasts the questionable nature of mattering. So we try to imagine a way of being, a source of meanings that goes beyond the daily life that hastens to an end, when our life shrinks from a world, an office, to a data point held, for a while, by the Office of National Statistics.

At first, eternal life was located in a hidden place, removed from space and time, from here and there, to be revealed after death had disrobed us of our perishable flesh, unmasked our face as a mere mask. In this place, ancestors were joined, friends reunited, God met face to face, and the true meaning of everything disclosed to our awestruck gaze. Most importantly, *change was arrested*. And there was the catch: unchanging reality could just as well be unremittingly horrible as permanently pleasant. This uncertainty, which has haunted humans whenever they thought of death, was exploited by those who wished to command their obedience this side of the grave. The 'hereafter' became, in the hands of those who were accredited with a privileged understanding of the will of God, an instrument to terrorize those who needed to be kept in order. Mesmerized by promises of eternal bliss and threats of unending punishment, men and women stayed in line and colluded in their own continuing subjection.

That, at least, is what the cynics tell us. But even those who feel that the idea of immortality is deeper than the this-worldly politics with which it is tainted find it troubling. At the very least, the notion of eternity, wobbling between time endlessly piled-up and time frozen, is obscure. Nor is the prospect of a purely spiritual, discarnate after-life very enticing, given that most of life's pleasures are mediated by the flesh. The warmth of the sun on our face, a satisfying belch, a kiss, not to speak of more intimate carnal delights, would seem to be ruled out. A headless, disembodied 'I', what is more, would have no location in space; and, since space and time are inseparable, this 'I' would also be unlocated in time: 'nowhen' as well as nowhere, it would have nothing to do, nothing to find out, nothing to achieve, nothing to think, nothing to be; little more than a wraith haunted by a diminishing stock of end-lessly replayed memories. Death might seem preferable to such a salvation.

The alternative – resurrection of the head – raises even more questions. How old should my restored head be? Spare me a rerun of my nappy-filling infancy. Spare me my teenage years and that gauche, half-remembered stranger who still makes my toes curl. Spare me an endless iteration of the afflicted body of my final illness. And what about those whom I hope to meet face to face? Will my mother be young or old? What if she had a say in the matter and chose to be with *her* mother who died before I was born? The Almighty would be hard put to reconcile all the demands of the Saved.

Other modes of fleshly continuation, for example Hindu recy-cling, seem both physically implausible and unattractive. It gives me little pleasure to think of my present existence as the after-life of a cockroach rewarded for moral excellence by promotion from an arthropod in a carapace to a human being in a suit. And what if I were punished for my sins this time round with downward bio-logical mobility? The thought that those sunny days I spent playing on the beach with my children might be re-experienced as a tingle of pre-existence by a frog commuting between lily-pads is not a happy one. And Nietzsche's Eternal Return, the inevitable result of the endless repetition of configurations of the atoms of the universe through the infinity of time, hardly inspires *amor fati*. This non-spiritual recurrence is, as he recognized, the most terrifying thought of all.

Little wonder, then, that people are increasingly seeking immortality, or proxies for immortality, *this* side of the grave. Not that we entirely believe that, by regular attendance at the gym, and ridding our bodies and our lives of cardiovascular risk factors, we can buck those thermodynamic laws which dictate the corruption of our flesh. However, the value of even modest postponements of our Absolute Undoing may be enhanced by the opportunity to buy fashionable running kit in which to exercise: such are the joys of mixing secular Calvinism with Calvin Klein. Combining Looking After My Heart with Looking Bloody Smart will go some way towards sheathing the spikes of panic. But it doesn't really solve the fundamental problem.

Not that anyone who really thought about it would want to prolong their present existence forever. The prospect of an infinity of days, even if they were healthy, is appalling. How long would the familiar continue to please and novelties bring delight? Just how many performances of *Don Giovanni* could we stand: a hundred, a thousand, a million? Would not life without termination lack shape, direction, even purpose? Do not the shadows of mortality enhance the beauty of ordinary daylight? And how would the first generation to have freehold (rather than mere leasehold) on their patch of matter cope with survivor guilt?

For many, the only attractive mode of this-worldly immortality is eternal fame. They are spurred by dreams of continuing in the consciousness of others; of being celebrated, anniversaried, admired, studied, printed on T-shirts; of having their Wednesdays minutely chronicled by biographers; or, more humbly, of living on in the gratitude of those whose lives they have made better by their passage through the world.

Even this fantasy withstands little examination. For a start, posthumous survival is not a function of goodness or any mark of worth. Hitler's memory will outlive that of millions of good people. More to the point, fame after death cannot be experienced – except as an idea in the mind of the not yet dead. If you don't enjoy being famous while you are alive it seems unlikely that you will enjoy it at all. Mozart today can no more savour his posterity than spend the royalties that would make him a trillionaire or enjoy those Mozart chocolates – sold everywhere in the Vienna where he had his wretched early death – on which the image of his bewigged head is endlessly replicated. His mass-printed cheeks rejoice in neither the sunlight that we, whose lives have been blessed by him, enjoy nor the adulation that bathes his memory.

Besides, the collective mind is hardly an accurate mirror of ourselves. Rilke spoke of fame in one's own lifetime as 'the sum of the misunderstandings gathering around a new head'.[21] And just imagine the misunderstandings that may gather around our names when we are dead. Think of the nonsense literary theorists have written about 'Shakespeare'. What kind of existence, moreover, does anyone enjoy in the minds of other people? Imagine being the grammatical subject of ungrammatical sentences scattered through the inner monologues of strangers – muttering to themselves about their promotion prospects, their weight, their chances of hooking up with the person opposite, and the train they are going to miss. I doubt that Shakespeare would recognize himself in the thoughts and conversations of those who adore him, never mind in the resentful soliloquies of those who are compelled to study his plays in order to pass an examination.

This seems a poor substitute for personal presence, or being self-present in the way that I am now. It is difficult not to agree with Woody Allen when he said that he didn't want to live on in the hearts and minds of others; he preferred to live on in his apartment. Bertrand Russell, too, reflected on this when he thought about the great Eleatic philosopher Zeno:

> In this capricious world, nothing is more capricious than posthumous fame. One of the most notable victims of posterity's lack of judgement is the Eleatic Zeno. Having invented four arguments, all immeasurably subtle and profound, the grossness of subsequent philosophers pronounced him to be a mere ingenious juggler, and his arguments to be one and all sophisms.[22]

On the matter of posthumous fame, we should allow Byron (speaking through his *alter ego* Don Juan) the last word:

> What is the end of Fame? 'tis but to fill
> A certain portion of uncertain paper...
> To have, when the original is dust,
> A name, a wretched picture, and worse bust.[23]

No CV, however impressive, will secure permanent tenure in the collective mind. Those we count on to remember us will themselves soon be dead and forgotten. It is doubtful whether even Shakespeare will be remembered in AD 40,000. Is an afterlife lasting a few millennia a significant advance on immediate descent into oblivion? It doesn't need a mathematician to point out that finite extensions of finitude still fall infinitely short of eternal life. And even those finite extensions are the playthings of chance. It was the fire that destroyed the palace of Knossos and baked the Linear B tablets that secured their scribes another 3,000 years of posthumous (if anonymous) existence.[24] The chaos that usually extinguishes our voices may be the medium that carries them.

Anyway, we can't all be famous. The dead compete with one another for the divided and distracted attention of the living. One person's posthumous glory requires the obscurity of thousands of fellow humans. If fame were apportioned equally, no one would be famous. Signals such as 'Shakespeare' would be lost in the noise. Every supernova dims established stars: even Zadie Smith steals a little of Homer's radiance.

No wonder the idea of immortality, this side of the grave or that, of the flesh or of the spirit, is itself marked for death, since it delivers little comfort that withstands reflection. The upward arc from speechless infancy to articulacy and downward to anecdotage is all we are going to get. We need to get used to the enormous, all-encompassing fact that all our purposes converge to purposelessness and all our mattering fades to insignificance; and that the world will soon efface the wake of our passage through it. How fortunate, then, that we are shallow, so that our dread of extinction is not as frequent, and rarely as intense, as our sexual desires, our longing for sleep, or our fear of embarrassment. And when the blessing of shallowness fails us, we might recall Nobel prize-winning Hungarian author and concentration-camp survivor Imre Kertész's claim that 'the smallest instant of life is stronger than the death that denies it',[25] and try to imagine this as if it were true.

In the meantime, before we bid farewell to our heads, we may meditate on the extraordinary creatures that we are; on how our individual worlds, the shared world and our heads relate to one another; and on the fact that our heads do what we have been doing throughout this book: thinking.

Final Explicitly Philosophical Digression:

Knowing (and Not Knowing) My Head

We have transformed everything by virtue of our unique human ability to collectivize the sentience that we share with other animals and make of it a property of a community of minds. I and the cat both feel warm; but only I experience *that it is warm* and explicitly share this with others as an acknowledged common condition. Our experiences have the habit of changing into facts. The consequences of this are endless. They could be summarized in this way: animals *live* their lives; humans *lead* them. We deal increasingly with facts, factual beliefs and factual opinions rather than responding to sensory stimuli – and this is as true of our dealings with our own heads as it is of our dealings with the world our heads move through.

You put on this lipstick because you know it suits you; because it is affordable; because it will last the full eight hours of your day and still be working when you go out in the evening. Because, because, because – journeys that mark a path through the infinite nexus of more or less general possibility that lies beyond the body's direct experience of warmth and cold, pain and pleasure, hunger and satiety. And the head itself, as we have noted, is the subject of a near-infinity of facts – more facts than the head could contain. This factual knowledge about our head is the most distant of our relations to it; the distance that makes this head of ours anyone's head, no one's head, a third-person head.

Knowledge begins with mediated experience. You see what your head looks like when you catch sight of it in the mirror. That woman, opening her mouth with a silly smirk on her face, and more wrinkled than she had realized – that woman is you. And then there are facts we learn or are told about our own heads. Your head is thirty-five years old; it weighs such and such; it is like your grandfather's head; it is affected by such and such a disease. And finally, there are facts that apply to all heads.

Knowledge is power. It extricates us from the immediate here and now; we are lifted up above direct collision with the natural world; we can approach the world as if from the outside. And yet it is also a source of a sense of helplessness. While knowledge is power, to be known is to be in the power of others. They can see what, in the broadest sense, I look like. I am in their keeping. And even those facts that do not place me in the power and judgement of others are a source of distress.

As the beginning of knowledge in the mirror suggests, the head that you spectate is in part an unfolding spectacle which you only partly animate with yourself. That silly, smirking face is not the you making those rather clever comments; and the information you have about your head – the truth about its secretions, for example – is far from you. No wonder you sometimes have a sense of being eaten away by all sorts of dull, impersonal truths that guide you through the great spaces of your life, that occupy your endless ruminations, and that describe the organic machinery that enables you to live, shapes your experiences, and even continues for a while when you are no longer.

This is the dismaying truth: knowledge of your head fails to be 'the low-down' on your head because it does not correspond to what it is like to be your head. The truth is, there is no 'what it is like to be my head'. There is no 'what it is like to be' *any* object of knowledge. This truth is prefigured when we look in the mirror and cannot fully connect ourself with that smirking, chatting face. And the master-fact in the realm of knowledge is that this relationship with the head that has much business of its own to transact is only temporary. The deepest bit of knowledge is that we shall die. This is what haunts the realm of facts; what the dissociation between what we feel and what we know and between what we know and what we don't know tells us.

You are and you are not your head. If you refrain from weeping at this thought, it may be because you know that your tears, rich in manganese, would dislodge the mucus in your nose, getting on with its own organic business. Let us instead enjoy Paul Valéry's beautiful observation in *Monsieur Teste*:

> We are made of many things that know nothing about us.
> And this is how we fail to know ourselves.

chapter fifteen
Head and World

And still they gazed, and still the wonder grew,
That one small head could carry all he knew.

Oliver Goldsmith, The Deserted Village

If this head were to try to encapsulate in a few words everything that is most amazing about itself, those few words would be: *it has a world*. And one of the most extraordinary aspects of this macrofact is that my head locates itself in that world. It is explicitly related to its world as its centre – in a sense that includes the frequent feeling that it is also on the margin, at the edge, 'a long way away'. My head, in short, is located in the world it contains within itself. The world my head inhabits is, in an important sense, located in my head.

My Head in the World

My head is in the world. Where in the world is it? It is here, where I am. And where is here? There are several equally acceptable answers: where this headache – the one I am currently suffering – is; in this room; in my house; in the road where my house is; in Manchester; in England; in the world. A linguist would note how, as the designation of my location becomes more broad-brush, it also becomes less 'egocentric'. I began with terms such as 'this' and 'here' and 'my', whose meaning depends on the person who is speaking and even on the literal location of his head. My voice exploits its own location. Then I moved on to proper names – 'Manchester', 'England' – whose meaning is independent of where I am when I use them. Unlike the meaning of 'here', the meaning of 'Cheshire' does not change when I move ten paces to the left. Unlike the meaning of 'I' or 'my' the meaning of 'Manchester' does not change with a change of voices.

Of course, the distinction is neither as clear-cut nor as simple as that. My Manchester will differ from your Manchester, my England from your England. Our Manchesters are built up out of ideas, experiences, and journeys. The gestalt psychologists talked about 'hodological space' – a space in which we locate ourselves marked out by the tracks we have made, by the customary routes we have taken; it is woven out of the 'ways' (the Greek for a way is *hodos*) we have followed over time. The poet spoke of 'an Italy of the mind', as if that were different from the real thing. But, of course, there are *only* Italys of the mind; or the sum total of such Italys built up in the great community of human minds. And this is true of my less romantic Manchester and England. I have 'my take' on Manchester. I live on the edge of Manchester, in a village called Bramhall, and that is where my head is most often located. There is something called the village centre, but for me, and for all but a few Bramhall residents, that centre is not usually 'here' but 'over there'. Besides no one lives in the exact centre. If it were to be agreed upon, it would probably correspond to a bollard at the main crossroads – not a very fashionable or comfortable address or one associated with a very long life expectancy. Bramhall is the sum total of everyone's 'takes' on Bramhall.

On the other hand, my sense of being here is not entirely lacking in objective frames of reference: 'Bramhall', 'Manchester', 'England' seep through the sides of my most absorbed moments, when I am head down, busy in the 'here' and 'there' that surround me. Only a small child can be lost in that 'nowhere without no' that Rilke spoke of.[1] The rest of us are inescapably mapped on a grid that locates us in a network of places we share with an infinity of others.

So, while the space in which my head is located is at any given time personal – shaped by the way I have mapped out the world in the busy, largely self-interested tracks I have taken through it – it is also impersonal, captured in maps, gazetteers and textbooks of astronomy. Space itself is both a given that locates me and a construct in which I locate myself; a matter of physics and of mental clothes, and of topography which shimmies effortlessly between the two. (That is how it is possible for love to transform a city, as happened to me, when Liverpool in July 1964 rearranged itself around the possibility-of-JS and I became the floating capital of her absence, with my head at its centre.)

There is nothing simple, therefore, about the notion of my head being in the world. This world is subject to incessant fluctuations: it expands, it contracts; it lightens, it darkens; it has a two-second horizon, it has a ten-year timeframe; it is one pair of eyes, it is endless plains; it is a feeling of warmth on my face, it is the adverse economic trends that are concerning us all; it is the location of a twitch, of a deliberate action, a tactic, a strategy; and so on. Nabokov took exception to the notion of writers representing *the* world. '*Whose* world?', he asked. Quite right, too. But I think we could go further: my head is located in a shimmering succession of countless different worlds.

It is not quite like that, of course. Behind the pell mell, there is a certain stability. That is why I can talk about it now. There is a world set out in physical space made up of something we are inclined to call 'matter', inert stuff that underpins all the volatile mattering that picks out this and that. Matter not only guarantees the coherence of my world; it also underwrites the coherence between my world and your world and anyone's world. Our heads glean different parts of what appears to be a literally coherent, or cohesive, world.

Somewhere between matter and mattering, between the nexus of material objects and events existing seemingly independently of us, and the tapestry of private significances, are 'objective facts'. In the opening sentences of his oracular *Tractatus Logico-Philosophicus* Wittgenstein asserted that:

1.   The world is all that is the case.
1.1 The world is the totality of facts, not of things.[2]

Facts are the struts in the lattice-work of the shared space of possibility – the world, the worlds, to which we relate as we move through our lives.

At first sight, this seems to make plain sense but nothing could be less plain than facts. What is a fact? It is not something like an object that is simply 'there'. If you doubt this, try counting the facts in the room in which your head is currently parked. You'll find the number depends on what you choose to notice and how you divide it up; and this in turn depends on how you characterize what is there. Facts are the progeny of a three-in-a-bed between my consciousness, my language (and the habits of noticing and dividing dictated by my language) and whatever is intrinsically there, independent of my awareness and my descriptive habits.

It will be obvious from what I have said that the way my head is in the world – more precisely is in *its* world and even more precisely, in its *moment*-world (a moment steeped in private and public history and pregnant with private and collective futurity) – is not the way one material object is inside another, larger one. The world is not a big bubble around the small bubble of my head, containing it. Though at any given time my head may be next to objects in the world – such as this cup of coffee – those objects and my head are part of the same world only by courtesy of my head, which brings them together in a unity. This brings us effortlessly to the other side of the glove.

My World in My Head

The reason my head is not located simply in the physical space it occupies is because it contains a world: it contains the world that contains it. We need to be careful when we say this. I don't for a moment wish to suggest that the daylit world in which my head is located has been cooked up in the darkness inside my skull. Indeed, the world in my head could be seen as being in part the sum total of all the worlds my head has found itself in. All of those journeys – supplemented by written, spoken and visual reports capturing the summed journeys of mankind – weave the world in my head. That is why the world in my head is in tune, and in communication with, the world in your head.

It is rather daunting to think of all those journeys; to recall the immense length of the world-line traced by my head over minutes, hours, days, weeks, months, years and decades as it has trawled through what is there and what is said to be there. What long pilgrimages are coiled up inside our heads. I look at my mother's head and think that it has stared at the sunlight outside the terraced houses in Edwardian Liverpool; that it walked along Scotland Road to work in the 1930s, burning-cheeked in a lather of shyness; that it listened in fear to the explosions of the May Blitz; that it perambulated one baby after another through the park; that it listened to my unbroken voice as it boasted of its achievements at school; that it wondered when her lively grandchild would ever fall asleep; that it cried with frustration as my ageing father became ever more impossible; that it looked at itself in the mirror and wondered who that old woman was.

No wonder 'here' is so complex; and no wonder our heads resolutely refuse to be confined to the irregular lump of space they fill or to the sensory field in which they locate themselves. And at any given time, we are not at that given time. Our moment-to-moment space occupancy is massively outsized by the hidden space of our preoccupations. The very sense we make of what we sense draws on a past acquired in a thousand nears and fars; and reaches into a future equally arrogant in its requisition of things that lie beyond horizons that lie beyond horizons. Each point in the long world-line we trace with our head through the shared world is in fact a huge packed sphere.

When I try to scope this world in my head, I feel a sense of defeat. Walt Whitman's claim – 'I contain multitudes' – seems too modest.[3] In an endeavour to triangulate I recall a few 'spots of time', flashbulb memories: Great-Aunt Nell, saying 'I believe so' in the kitchen c. 1953, sunlight italicizing the grizzle of her chin; dancing home from school after receiving a top mark; an afternoon when the world seemed like a painted screen and I started falling towards what I feared was madness; a pause in a summer field, listening to a poplar tree, swaying in a breeze that had orchestrated a county, knowing that I was in love; the terror of a patient fading away before me; the joy of an exchanged smile with our child; an hour in Amsterdam when I sat on a bench and felt as if I were grasping my thoughts for the first time; the delight of applause at the end of a lecture that has gone well; a silly remark that brings me to a halt in the street twenty years later.

Alternatively I can think of all the places I have sat, stood, walked, talked, gawped, slept, worried and delighted in during the last sixty years, encompassing four and a half continents, forty countries, a thousand towns. Or recall the hundreds and thousands of heads with whom I have interacted as a lover, friend, neighbour, colleague, doctor, auditor, fellow queue-er or adjacent mill-er in the same crowd. Then there are all the facts that furnish my world: the date of Parmenides' death, the diagnostic features of cardiac failure, the capital of Bulgaria, my wife's favourite novels, the influence of Hermann Broch on Milan Kundera. In addition, there are the thousands of items and modes of know-how that I mobilize in everyday life, and the dozens of acquired attitudes to people, places, things, and a myriad of abstractions such as Gordon Brown's foreign policy.

What things this head – the *tête* or *testa* – will have seen, heard, felt, known between the moment it popped out of the birth canal to the sound of its mother's cries of pain and joy and the moment when the crematorium doors slid behind it to the sound of music and sobs lie beyond any kind of summary: squinches, daffodils, stretches of the obvious, Wednesdays, good prose, unbearable pain, economic trends, stooped gaits, the possibility of seeing the Queen, intense irritation, beautiful faces, orgasms, zebras, the sense of being late, Renaissance polyphony, taxis, autumn leaves, saliva, faience, the light of dawn, the chaffinch's song, the variable duration of the chaffinch's song, dashed hopes, teasing, sounds muffled by distance, inverted ticks, lemmas, road humps, anticipation, fossicking, successful grant applications, stepping stones, arguments for evolution, potholes, staples, windsocks, astonishment, myths of flight, distant mist, bloom on plums, surging prices, catalepsy, flare-ups of illness, pipe-cleaners, the Advertising Standards Authority, platforms, Poetry Consultants to the Library of Congress, kerfuffles, forks, mathematical puzzles, filter coffee, trousers in and out of fashion, instances of a lazy man's load, toenails, lighthouses, moral judgements, pranks too far, clouds, coat-hangers, sand, solitude (longed- for), solitude (feared), washing powder, drubbings, nuances, assonance, a dearth of good garden architecture, bird-fancier's lung, vaginal secretions, cantatas, bollards, stagecraft, cricket, puddles, hydrangea, chance-met strangers, souped-up cars, coincidences, space, happiness, ducking, the scent of sandalwood, streetlamps like apricots, tares, regulatory bodies, fading fabrics, empty laughter, lost things, salt spoons, offences not taken, bans (outright), bans (partial), bolted breakfasts, glad-handing, boredom (hours of), dust, resorts, clinics, persecution of Cape Coloureds, steeples, stooges, radio interviewers, climates of opinion, holidays by the sea, pylons, competing theories of motorway signage, tittering, periods of my life under an interdict against tittering, the ticking of boxes and clocks and in mattresses, political parties that lack pizzazz, finbacks, drafts in pubs, dry stone walls, sneaks, quirky senses of humour, humans who might conjecture or worry that every polynomial or real number that never takes on a real value might be representable as a sum of squares of ratios of polynomials, Atomic Energy Authorities, more pranks, affidavits, menacing glances, award-winning footage, throwaway remarks, elastic bands, welters, a fondness for frequentatives, tennis rackets, showers, eastern promise, lovers of lists, inappropriate reification, lilies, terrifying glimpses of nothingness, courtesy, photo-opportunities, faceless bureaucrats, wonder at medieval tapestries, disputed attributions, cartilage, squirrels, traffic cones, twists in the tale, undeserved acclaim, crows, baseball...and so on. The fact that this list has been compiled in places as remote as Hong Kong, Prague, Kyrenia and Bramhall just adds to the wonder that so much can be stored in my head and kept in its right place.

Of course, it is not all there, up front. But it is available and it continuously draws us away from the thistledown lightness of the moment of experience – except when pain or terror reduces us to a frozen instant. And so I cannot but wonder how this world is accommodated in this head; how the past is 'stored' and a highly structured future is planned and prepared for and anticipated in a place that, in common with all other pieces of matter, has only a present tense – or indeed (from the point of view of physical science) no tense at all. How that smile is recalled from 1970, and is allocated to its own world, connected with a boundless nexus of the circumstances in which it was given with such love and received with such delight.

Answering such questions has led many thinkers – probably the majority of those who have expressed a view on the matter – to imagine that the world in the head is a model of sorts and that the model is stored in computational form. I, or my head, or my brain, or something, is a kind of digital computer in which my past experience, my memories, my knowledge, my habits, my skills, my acquired attitudes are stored as bits and pixels, as patterns of nerve impulses. This would be fine if there were not the small detail that we are aware of the world in our head and it permeates our awareness of the world before us, so that we can make sense of the latter. Computers, however, are not aware in this way. This is no minor difference. It makes computers nearer to pebbles than we are to computers.

The need to add on awareness is easy to overlook because we tend to cheat by attributing to our computer-head or computer-brain activities such as 'information processing' and 'representation' and 'modelling' which, while they are also said to be carried out by (unconscious) computers, actually seem to be shot through with awareness. As a matter of fact, computers don't inform, represent or model anything in the absence of conscious beings; just as clouds aren't signs of rain in the absence of creatures that are aware of the world around them.[4]

Future generations will look back in amusement at the way we were so impressed by our own technology that we modelled ourselves on it – that we thought the world in our head was a computational mock-up built out of neural activity. They will be even more amazed that we imagined that we could build up our own subjectivity out of the activity of one wet *object*, the brain;

that we imagined the inquiring subject could be itself a transparent object of our understanding; that we might gather up in a few pieces of knowledge the explanation of the consciousness out of which knowledge has grown.

The world in my head is very susceptible to change. For a start, the inner world that is engaging with the world that surrounds me is only a minute sample of what is held in store. Secondly, events can change that world irrevocably. A mere bang on my head may make a massive dent in my world, bits may disappear, its inner struts fall apart, and my ability to access the remainder when required may be seriously affected. I may be world-lean, to the point where I may stare at pages, towns, streets, the faces of those whom I love or even the back of my hand with incomprehension as I fail to link the present moment with the past, and the nimbus of sense no longer glows from the objects that are before me.

But there are other more cheerful ways of transforming the world in my head: for example a pint of Stella Artois, which I can pour through that vent in the front of my head and, lo and behold, the world glows with a new light, and the head is differently attuned to the world without. Suddenly the past seems more profound; the future with its burdensome demands and potential ordeals lighter; music sends warmer evening light across my soul; and the smile on the face of that man over there seems more beautiful. Thomas de Quincey felt this miracle when he discovered laudanum (though much sorrow was to follow):

> Here was a panacea… for all human woes: here was the secret
> of happiness about which philosophers had disputed for so many
> ages, at once discovered: happiness might now be bought for a
> penny, and carried in the waistcoat pocket: portable ecstasies
> might be corked up in a pint bottle: and peace of mind could be
> sent down in gallons by the mail coach.[5]

Oh that the relationship between the head and its world, between the head around the world and the world inside the head, were so malleable. And that the microcosm within and the macrocosm without could be as harmoniously related as Pierre felt when, in love with Natasha, he walked home under a clear, frosty night sky and saw the brilliant, portentous comet of 1812:

> This heavenly body seemed perfectly attuned to Pierre's newly melted heart, as it gathered reassurance and blossomed into new life.[6]

And yet without such peace of mind to help us withstand the knowledge that the world is largely indifferent to us, that we are but a minute part of what we know, we would have been so much less.

Each Head in Another's Head

> One person is always in the head of the other,
> and this head in turn is in still other heads.
> Friedrich Nietzsche, Daybreak

The world in which my head is deployed interacts in a multitude of ways with the world inside my head. The result is an extraordinary mixture of propositional thought, random behaviour and deliberate action regulated by practical reason, confusion and clarity; and of intangibles such as sensations and feelings and memories, flickering over reefs of solidity – my career, responsibilities, inalienable rights, enduring relationships and possessed possessions. I am an evanescent lacework of lived time enclosed in a many-layered crust of curriculum vitae, which defines my duties and rights in the world. I would like to say that I wobble between a life passively experienced and a life actively led but that would simplify matters to such a degree as to undermine the point of this chapter. Better to say that the two modes of living are inseparable – like those rainbows pitched on the cataclysm of a sunlit waterfall, their miraculous stillness staining the mad mist thrown off permanently hurrying water.

The degree of control evident in my life – such that, for the most part, I seem to lead my life rather than merely endure my body – is impressive. Not infrequently, I will set out early in the morning and arrive, several hours later, at a destination hundreds of miles away, precisely on time and equally precisely in the right place. The miracle of an ordered life actively led through a medium of uninvited happenings passively received, amidst digressions consented to and digressions from digressions, is something that I mostly fail to notice. I take the honeycomb for granted and overlook the million flights, the patterns of bee-lines that wove it so densely. That is why I treasure those few moments when I can remember the stuff out of which I – this upright, middle-aged, articulate, serious man – am made of.

Such a moment occurred a few years ago, when I was waiting for a plane taking me to a conference where I was going to talk about epilepsy and other interruptions of consciousness. I saw 'Florence' flashed up and was suddenly reminded of Florence S., one of the twenty-four patients I had seen on the ward round after the night my team was on duty for emergencies. I had forgotten to refer her to the neurologists for an urgent opinion. This moment introduced me to the notion of 'cognitive luck' – analogous to moral luck – whereby we are reminded by accident of things we should have remembered to do. And now I see this force at work all the time. It was a moment when I woke out of daydreaming to the associative seethe of self-presence, which has somehow to be shaped into character, role, office so that when I greet and meet, I know whom it is I greet and what I meet.

Short-term, semantic, procedural, episodic and autobiographical memory all work together so that I can remember the last sign I drove past, understand why it was put there, know how to drive the car, recall the last time I went to the place I am driving to and how I felt when I went there. And when I have a momentary lapse, I rack my brains and somehow shake out the relevant fact, as if I knew what I was looking for all the time and the cues I should use to flush it out.

At any rate, the random events of the outside world seem hardly the comb to untousle us. Though the outside seems much of the time to be a mass of distractions we have to surf to reach the shore of a completed task, we seem for the most part calmly at home in the world. How do we accomplish this? One answer is rather amazing: we are almost masters of the dangerous art of making generalizations on the basis of very few data. And when we catch ourselves in action – making such presumptions – it is rather extraordinary to see the kinds of prejudices by which we steer our way through the world. Consider, for example, what happens when a middle-aged woman nips in ahead of you and steals the parking slot you thought had your name on it or a youth in a fast car cuts you up, causing you to brake sharply. How promptly your anger makes sense of the woman's heavily made-up face and blonde hair, or the man's shaven head, his tattooed forearm, and the turbo-charged car throbbing with loud music! The Venn diagram of your biographies shows the slimmest ellipse of overlap. Nevertheless, using only minute samples of their lives, you seem to have their number. It is obvious that the woman votes Tory, supports capital punishment, is a parasitic lady who lunches, and reads the *Daily Mail*. As for the youth, it is equally self-evident that he did not arrive honestly or usefully by the wealth that enabled him to buy that car, that he treats women with contempt, that he is the kind of moronic, conscienceless yuppie who constitutes much of what is rotten with present-day Britain. Thus do we 'make the world our thing' (to vary Levi-Strauss).

There is nothing special about these toe-curling examples. All judgements are, and have to be, more or less snap. There is scarcely a pause between experience and classification: the moment-to-moment business of consciousness is a succession of leaps of extrapolation. Even so, it seems at times that it is a kind of private madness that sustains our shared sanity; that civic society is founded upon a delirium of necessary prejudices which enable us to live in a knowing state in the world. Without such presumptions and generalizations, we would be in trouble. Daily life would indeed be 'a torment' as Leszek Kolakowski has described it, with 'no link…between disparate events' so that 'nothing is real, nothing is really experienced and everything dissolves into a chaotic mass of details'.[7]

chapter sixteen

Thinking Head

It is no accident that I have left this investigation of the thinking head to the last. Throughout this book I have been trying to get my head around my head. And now I am about to try to get my head around the very processes by which I have been trying to get my head around itself: I am going to think about thought. The vertigo that beckoned when I looked at myself in the mirror so many pages ago threatens to return: this is my thinking head thinking about the thinking of my thinking head. No wonder that, when I try to focus on this matter, my head seems to empty in advance of my thoughts – indeed to be hollowed out by my thoughts – as I feel the pull of the whirlpool of reflexivity.

But I exaggerate a little. This process is not quite like the impossibility portrayed in the Escher pictures of the hands drawing themselves into being. For the collective of heads past and present provides a mirror in which my thoughts about my own head and about my own thoughts may seek their image. Even so, this is no place for the faint-headed.

Thoughts occupy my head all of the day and much of the night. Worrying, ruminating, hoping, loving, hating, planning, recollecting etc are accompanied by, specified in, even expressed as, an endless flow of more or less bad prose. While the stream of consciousness is not just a stream of words, we are persuaded by novelists of genius such as James Joyce that wakefulness and wordiness are almost coterminous.[1] This head of mine is a locked tower of sentences, unlocked by the tongue or more slowly by my fingers on the keyboard that are typing out this sentence.

Let me resort to my favourite prop: the mirror, in this case, the spoon with which I have just finished scooping up ice cream. I look at my face and my face returns my gaze. There is nothing in it that could tell me that I was thinking, even less what I was thinking about. Of course, it is a safe bet that I am thinking. That is why I can guess that that lady over there – my wife of thirty or more years – is also thinking, though she is sitting in silence. And as I look across the sunlit square, crowded with brilliantly sunlit heads, I know that those who are not talking are thinking to themselves: there is a chattering darkness in each of those dazzled heads, pointing to places outside of the sunlight, grazing over distant semantic fields, rummaging in the private nooks and crannies of the shared space of possibility.

I close my eyes, so that I can redirect my gaze inwards. The dark, endless night of the skull, interminably nagged by my voice. I think about the thoughts that grow in that darkness, the thoughts I cannot avoid. I cannot but hear my thoughts; yet I cannot hear yours. The decades we have spent together have made your thoughts no more audible. When, as I did from time to time, I applied a stethoscope to a patient's skull, I sometimes heard the bruit of the angioma I was listening for but never even the slightest rumour of the thoughts that I knew were ceaselessly passing through his mind.

It is worth thinking a little about the privileged access we have to our own thoughts. I may not remember what I thought a minute ago and cannot predict with certainty what I will be thinking about in a minute's time; but I know that, and what, I am thinking at this present moment. No one else can know this with any kind of certainty, though they may guess. This is why we tend to think of ourselves – and in particular our heads – as a kind of private space in which special – secret, occult, mental – events take place. This was dismissed by many philosophers of the last century as a false way of thinking about thoughts and selves. It is the result, they argued, of extending the idea of space from the physical world where it rightly belongs, to a putative mental realm, where it most certainly does not. Perhaps so; but we have no better way of capturing this private territory in which thoughts take place. We shall come back to this when we think about whether thoughts have a location.

First, however, let us address a couple of very obvious questions about thoughts. In what sense do we *hear* these events that no one else can hear? And why do they have to be heard by the thinker to be thought? As I listen to my thoughts just now, they seem to have an explanatory tone of voice. (This may be to reassure myself that an explanation is in the offing.) On other occasions, they sound kind or angry or sarcastic, as I rehearse various scenarios. Different thoughts, it appears, have different tones of voice. So, although they are silent, they seem to utilize dialects of silence. At any rate, there is a real sense in which they are audible – to me at any rate. Just how real is demonstrated by patients with psychotic illnesses who interpret their own thoughts (which they may feel have been 'inserted' into their minds) as the voices of others – mocking, instructing and barracking them.

The audibility of thoughts that no one else could possibly hear is a delicious embarrassment for psychologists and philosophers, particularly as thoughts lack other features of things that are audible. They do not have an obvious source – apart from me (whatever 'me' means here) – and consequently do not originate from a particular direction at a particular distance. And I do not have to strain to hear them. When I pleaded with the children to be less noisy, as I could 'not hear myself think', I was speaking consciously metaphorically. More interesting is the question as to why I have to hear my thoughts in the first place. Do I have to tell myself my thoughts in order to think them? In which case, how do I know what to tell myself, if I haven't thought my thoughts? Everyone knows the old joke 'How do I know what I think until I see what I say?; but nobody knows quite how to deal with it.[2] It raises all sorts of questions about the relationship between events and actions, passivity and activity, in something – thinking – that which seems to be closest to us. It causes us to examine the verb 'to think' and the action of thinking.

At any rate, it is clear that thoughts have somehow to be individuated and the way we are used to doing this is by articulating them in the language we have in common with our fellow men. The most explicit way of having a thought, therefore, is creating a sentence, even though we do not craft those sentences with any great care and often settle for fragments. What is more, having our thoughts as sentences – even if there are various features of sentences, such as tones of voice, which are not relevant to the core of the thought – seems to be the only way of having the thought. To have a thought is to hear ourselves thinking it.[3]

Are my Thoughts in my Head?

Earwax is in my head. Mucus is in my head. My brain is in my head. But are my thoughts in my head? We have already been skirting around this problem, as we have been thinking about the silent darkness in which our thoughts take place. Let us approach it from a different angle.

Philosophers, who seem to have the capacity to doubt things that no one would dream of doubting, sometimes wonder whether we really do know our own thoughts. Yes, we know that we are thinking, they say. When Descartes thought he had brought doubt to an end by saying 'I think therefore I am', he chose 'I think' as his starting place because, he pointed out, no one could be mistaken that he was thinking. The very mistake would be a thought and thus not a mistake at all. 'I am thinking' is a self-affirming thought. The question, however, is whether we know *what* it is we are thinking.

Of course, I know what I am thinking *about* but this is not quite the same. The only way we can know our thoughts is by listening to the sentences in our head. These sentences are made of words and the meanings of these words are determined by things that go beyond our thoughts. I may be having a thought about Aristotle and imagine I am thinking about a man who was tutor to Alexander the Great. A scholar suddenly discovers that Aristotle lied on his CV and he never tutored Alexander the Great. I, who thought that in thinking about Aristotle I was thinking about Alexander's tutor, now realize that I did not know what I was thinking about.

Arguments like this – and more subtle ones involving twin earths where water is actually 'twater', which has identical properties to water but a different atomic composition – have exercised philosophers for over thirty years. They have supported ideas such as that we do not know our own thoughts and that the meanings of thoughts, or of sentences that make sense to us, 'ain't in the head'.[4] The claim that we do not know our own thoughts is based on a fusion of the correct notion that the language in which we formulate our thoughts is not entirely transparent to us, so that we may use words mistakenly, and the incorrect notion that what I think I think is not what I think. The more radical claim that the meanings of items such as other people's utterances and our own thoughts are not in our heads is the result of confusing the meaning something *should* mean to us (if we used the words correctly or the world was as we thought it was) and the meaning something actually *has* to us. Because the former is not in our head, it does not follow that the latter is.

There is another reason, however, for denying that my thoughts, and the meanings they have, are in my head. This is the venerable one that says that meanings and thoughts are not the kinds of things that occupy space. To suggest that they do is, according to the thousands of philosophers influenced by the Oxford don, Gilbert Ryle, to make a category error: to talk about the location of thoughts is like talking about the nutritional value of prime numbers.[5] This kind of thinking is easier to challenge if one separates the notion of a thought as something that marks a general position in logical space from the notion of a thought as something that happens, that occurs to me, as something that I have. When we think of the latter, then the category mistake argument looks distinctly weaker.

Supposing I am at Stockport station. It seems reasonable to think that the thoughts I am having are being had at Stockport station, just as it seems reasonable to assume that the sensations I am having are being had at Stockport station, given that that is where my body is. As regards sensations, it is easier to see this with some sensations rather than another. Consider an itch on my arm. I have this itch in the very place where the mosquito landed and bit me and this, being at Platform 1 on Stockport station, was a certain distance from the buffet also on that platform. In the case of visual experiences, it is perhaps less easy. When I see the train, the sight is not in my eyes; it as it were refers itself to the object that is seen: the train itself. The question we have to ask ourselves is whether a thought is more like a visual image – of which it does not make easy sense to say that it was located at a particular location – or an itch, which certainly invites the observation that it occurs at a particular place. Thoughts are, it seems to me, more like itches, not because they do not refer beyond themselves but because the things they refer to are typically not present. When I think about you, my thought is not commingled with, inseparable from you. Indeed, my thought, explicitly about you, is explicitly separate from you.

There are two other reasons for thinking of my thoughts being co-located with my head. The first is that the thought occurs in time: I had such and such a thought at Stockport station. I was at Stockport station from 2 until 2.30. My thought must have occurred at some time between those two limits. Now, whatever occurs in time must also occur in space: it must take place some-where if it takes place some-when. The only plausible location in Stockport station is my body; and the only plausible place is in my head. Other body parts – my leg, my spleen and my toenails – seem less plausible. Even less promising candidates are the head of the man standing next to me, the luggage trolley, the air between us, or the space between the molecules of the air. The other reason relates to a very common form of thought which philosophers have called 'perceptual-demonstrative' thoughts.

Consider the thought 'That porter is very helpful'. The object the thought is about is the object perceived, over there, a couple of yards from my body. The reference of a thought that contains words like 'this' and 'that' (so-called indexicals) depends on the physical location of (the body of) the thinker. If the porter were not in sight, I would not know what I was thinking about and what the 'that' in 'that porter' picked out.[6]

The rediscovery that thoughts are *in* the thinking head has had the unfortunate consequence that many philosophers have identi-fied thoughts with activity in the brain of the thinker. A thought, they say, is simply a cluster of neural activity. This conclusion is not valid. The fact that thoughts are in the head does not mean that they belong to an isolated part of an isolated brain. The thinking thinker is a person open to the outside world, aware of his past and future, belonging to a community of minds and consciously located in that multilayered 'here', as we noted in the last chapter. None of these things applies to isolated neural activity. So perhaps we need to be a little careful when we think of thoughts as being in our heads: we will end up thinking that they are in our brains; or in tiny bits of our brains. We will then start imagining even that the thought that my thought is a collection of nerve impulses is itself a collection of nerves impulses. That is one bit of rather sophisticated biochemistry.

We can't help thinking in this literal-minded way, though. 'A thought just popped into my head' I say. We bang our heads with our knuckles to jog our memories and we close or screw up our eyes when we are thinking hard, as if to tilt the rivalry between outer and inner light in favour of the latter, as when we draw the curtains on a bright day to see a slide show more clearly. I have a friend who fell victim to a desperate depressive illness in which he had thoughts 'he could not get out of his head'. He used to scratch his head endlessly, leaving his scalp bleeding, as if he were trying to eradicate the thoughts inside it. And I myself went through a period in which I became helplessly preoccupied with some rather glum thoughts. I found myself shaking my head, as if I were literally attempting to dislodge them, like getting water out of my auditory canal after swimming. We should have known how hopeless this was, because the thoughts that made us suffer also spilled out of our heads into our fast-beating hearts, sweating palms, sickly, loose-feeling guts, showing their propositions as fastened to attitudes that fill our body-world.

The notion that my thoughts are so completely in my head as to be identical with nerve impulses in a part of my brain seems to have a lot going for it – so long as we don't clearly distinguish the two faces of a thought. A thought in one sense is an occurrence that takes place where I am. This is evident from those thoughts that rely for their meaning on my location – the 'perceptual-demonstrative' thoughts I have spoken of already, like 'That litter is a disgrace', where the reference of 'that litter' depends on where I am. On the other hand, the meaning of terms like 'that' and 'litter' and 'disgrace' belongs to a system of signs that is not located in any of us or in anywhere.

It is this Janus-face of thoughts that makes them so puzzling. Thoughts have two aspects: the events of my thinking them ('token thoughts') and the general, intelligible thought they exemplify ('thought-types'). This distinction is easier to understand if we think of a written word: the instance of the word 'word' which you are just reading now is a token and it is an example of the type 'word' which can be used again and again. Thoughts as tokens belong to individual heads in a rather literal sense, but thoughts as types belong to the network of symbols, the nexus of meanings, the space of possibility, created by the collective of heads.

By now, reader, you may be fed up with these technical philosophical arguments and in this you would be justified, if they were merely technical. Philosophers have the habit of concealing the profundity of the issues they deal with under what look like merely professional arguments and in-house quarrels. This is no exception. For what we are dealing with here reaches to the bottom of our condition as human beings, as embodied subjects: our elaborate joy and our ultimate dissatisfaction; the *splendeurs et misères* of encephalic life. In thoughts that are at once here and within us like sensations and elsewhere, and nowhere, we see our divided consciousness exhibited with especial clarity. That is why I want to think a little bit more about the products of the thinking head.

More Thoughts about Thoughts

I have just made some pretty strong claims about thought, the nature of human consciousness, and the dividedness of the human condition. I would like to explore them a little more because they lie at the heart of the preoccupations that have provoked this book.

A thought is a supreme example of an item of consciousness that reaches beyond itself: it is *about* something other than it is. This is true, of course, of perceptions: the object that I see and the seeing are distinct. I am aware of this distinction – when I see that what I see has a back, an interior, an underneath, that is concealed from me. In the case of the thought, however, nothing is revealed. That which it is about is entirely hidden. When, for example, I think about someone who is out of town, or about a holiday destination, or about the battle of Waterloo, or about the future, these objects are not exposed to me. Indeed, what I think about is incompletely specified. The person that I think about – however intensely – is vague, blurred, generalized. The person is present as a possibility or a bubbling magma of possibilities. The objects of thought belong to the space of possibility and vastly expand it.

This space of possibility, only faintly and indirectly tethered to actuality, eats away at us: it is a postulated presence that is in fact absent. As (incessant) thinkers, we are gnawed by absence, haunted by ideas that do not correspond precisely to anything we might experience when it is present. We have a new kind of view and our thinking head is a new kind of viewpoint: a random point of origin in an infinite ocean of that which is the case.

This is how I justify what I said towards the end of the last section. I come upon the same thoughts at intervals throughout this book when I remark on the gap between the facts about our head – the things that we or others know – and the immediate experiences of our head. The felt difference between the tears prompted by ocular pain and tears prompted by sadness is not captured in the objective fact that the latter are richer in manganese than the former.

This is in part the reason why the thinking head feels itself to be empty rather than full; why thought is existentially rather thin. It is not the entire reason, however, why thought feels slippery. I don't know about you but my most frequent thought since I started thinking deliberately about things is that I can't think. This is sometimes because thought makes too great a demand upon us: we think that there are things that should be thought that we cannot think, that there are thoughts that should come which simply fail to come; that there are connections waiting to be made that elude us. But it goes deeper than that: the problem extends beyond 'difficult' thoughts and 'difficult' topics to any kind of thought.

A cursory glance at the phenomenology of actual thinking shows it as a kind of chasing after itself, a slithering and a sliding towards and away from the object of thought, and so from the thought itself. What occupies us when we think is often a succession of fugitive ill-formed sentences distantly corresponding to even more elusive impressions, preoccupations and intuitions. This is true of thought at its most active, reflection at its most self-directed. For most of our waking lives, however, our thinking is almost passive: the 'I' that thinks is as much a *site* of mental events as the initiator of them; a conduit for fragments of language connected by associations of ideas as much as by the rules of inference. Our thoughts are often given to us rather than generated by us: we find ourselves thinking about such and such; we suffer our thoughts as much as we enact them. While thoughts seem to us to be entirely mental, and to come from 'within', they do not seem to lie clearly and exclusively within the domain of our discretion.

When we first encounter a transcript of the mind of *Homo cogitans*, we experience a sense of shock as well as recognition. It took the genius of James Joyce to acknowledge what it is like 'in there' – when he captured the stream of consciousness of mature, modern humanity, in so far as it is articulated, in the form of an interior monologue. This humanity appears like a blizzard of scarcely coherent particles of cognition. Ordinary thinking man, it seems, is light as thistledown. This was true as soon as we began thinking; before written words, telephones, televisions, emails, text messages, 'e-ttenuated' us even further and our essence thinned to e-sense.

The idea of being an agent in control of your thoughts is not an easy concept. The least requirement is thematic control, as opposed to a mere sequence of associated ideas whose extreme expression is delirium. But more than that is required, if we are to seem to be entirely in charge. The thinker, to be truly thinking his thoughts, should be, as it were, standing on the chosen spot, like a kestrel at stoop. This may not, however, go far enough to satisfy the notion of truly thinking a thought, which consists not merely of cognitive journeying to, and beyond, a conclusion but arriving at, and staying with, that conclusion; concluding with the conclusion without stopping thinking it. What would satisfy such a notion?

Stasis is difficult to achieve in thought. For thought is intrinsically mobile: its being lies in becoming. This prevents a thought from growing to fill the contours of the thinking self, or the self from focusing to conform to the contours of a thought. We cannot seem to possess, to have ownership, of our own thoughts because they, and ourselves, are multiple and our relationship to them – the relationship between the successive moments of the self and the successive thoughts that we entertain or are engaged by – seem contingent. The thoughts that I am presently thinking seem the products of accidents of outer and inner circumstances, the local inner and outer histories. We cannot seem to adhere to our thoughts nor do our thoughts adhere to us: their coming and going seems like lightning floods of words which do not stay with us and in which we cannot abide. There is no point of arrival where we and our thoughts coincide.

No wonder thoughts seem ontologically and existentially thin and we, as thinkers, ditto. Thinking is somehow hollow; any presence it has is undermined by elsewhere: its objects, the system of signification to which it belongs, its antecedents and its consequences. What little substance it does have – images, silent soliloquy – is somehow not of its essence. It does not deliver what thought is or aspires to: it has the character of mere supporting material or even a contaminant. Thought lacks the intensity of physical pain or pleasure, of toothache or orgasm; it lacks the commanding force even of fatigue or sleepiness.

What amounts to a slight dissatisfaction with thinking in everyday life, therefore, becomes an unappeasable hunger in philosophy, where the (rarely stated) aim is arrival in a thought that one inhabits, so that the river of cognition slows and swells and finally halts in a lagoon that has the shape of the thinker, of intellectual plenitude. (Under such circumstances, it would be as true to say that 'The thinker was thought by the thought' as that 'The thought was thought by the thinker'.)

There are other grounds for anticipating that the quest for luminous cognitive repleteness will be doomed. For example, there is no reason why the entirely satisfactory account of the world of the thinker should coincide with, or include as part of a seamless whole an entirely satisfactory account of the place of the thinker in the world. Nor is there any reason to expect that the thought corresponding to this account, if mind-portable, should gradually permeate, indeed assimilate, all the tissues of the self and mind, and should seal itself off both to new, irrelevant contingent events; that such a thought will become a perfect minding or self-reminding. Thought somehow fails to fill the territory it demarcates and thinking has no guaranteed point of arrival in a single intuition that can be held in a single moment of the mind.

Nor is there any point at which I can feel that I have thought what thought is or, more to our present point, got my head around my head. Every time I think about my head, it seems to disappear from immediate awareness and becomes scattered across a thousand places in the space of possibility, a latticework of facts. It thins to absence and, if it returns, it does so in the form of de-experiences, those cognitive ghosts that are bits of knowledge.

On this slightly defeatist note, my exploration of my head, that strange zeugma of intelligence and meat, of thought and bone, comes to a halt. The head itself seems hollowed out, transparent, weightless. Until I pop a mint in my mouth and the ghost robes itself once more in flesh.

For a little while longer, anyway.

Epilogue:
Heading Off

It is time to say farewell – but because there is so much unfinished business this is sure to be *au revoir* rather than goodbye. Let me end on a note of praise, a much-needed corrective to the current fashion of denigrating humanity.

The word is out in academe and more broadly among intellectuals that human beings are really beasts – and rather nasty ones at that – or zombies.[1] Those who take this view think that we are revealed, even defined, by the worst things we do to each other – by domestic cruelty, aggression, selfishness, wars, oppression, genocide and so on. Those who view us as zombies argue that we are wired in, via our evolved brains, to the outside world in such a way as to ensure replication of our genetic material. We are disposable phenotypes, utilized by the genome to ensure its own survival. Consciousness is well-nigh superfluous – mechanisms do the business for us – and free will is an illusion: our every action is the outcome of causes that lie outside the scope of our intentions. A recent magisterial history of the twentieth century – *War of the World. History's Age of Hatred* – has argued that the racism that drove so many of the horrors witnessed in that century can be explained by the evolutionary psychology of the human race.[2] And

John Gray, in his *Straw Dogs*, describes man as *H. rapiens* and a plague on the face of the earth, whose dream of progress has cost the planet dear.[3]

I am not too sure what we are supposed to do with these claims: if they are true, then, being beasts or zombies, we will not hearken to them; and if they are untrue, then they constitute an unwarranted attack on every source of hope, an act of supreme intellectual wickedness. If such misanthropy commanded belief it would not affect comfortably placed beasts and zombies – such as John Gray, the Professor of European Thought at the London School of Economics – while it would damage the prospects of those who are currently without hope.

In fact, our achievements – in creating civilizations, in constructing great artefact-scapes such as cities, in making an increasingly extensive and yet coherent, and certainly powerful sense of the world; and in regulating our behaviour towards one another (with terrible lapses) – are astounding. They are even more so when we consider the material with which we are doing this; when we think of the sleepy, confused, blurred, chuckleheads that were given to humanity when the first hominids woke to themselves and to self-consciousness. We were never given self-transparent souls, implanted from a transcendent source: we have created our own transcendence and, notwithstanding the curdled material out of which it has grown, have increasingly collectively liberated our heads from the biological destiny for which they were fashioned. Out of the coming-together of our heads, we have created social facts as robust as things, cultural constraints that pass themselves off as laws of nature.

Imagine a creature that became conscious of its own body *as* its body. What a medley of sensations would greet it; what a variety of things of which it might think 'This is I' or 'I am feeling this' or 'This is how it feels to be me'. These would have to be gathered into a sense of a self unfolding from moment to moment. We have seen what a heterogeneous mixture of things they are, as viewed from the comparatively solid or stable, established self. Imagine, then, how it would look or feel when the self is being established. Imagine that self then intuiting other's selves – projecting into others that unified sense that it has of itself. Just how great these achievements are becomes evident when we encounter humans in whom they are missing: those who are autistic, who have no coordinated or integrated sense of self and no sense of others as being independent, enduring points of view.[4]

On top of all this, we then have to acquire a sense of the independence of the world and its infinite lattice of 'that which is the case'. By this route, the sense of one's self (at least in part structured by an internalization of the General Other) and of other selves and of the world is pooled into the evolving cognitive heritage of mankind. That it should evolve – given that it is disconnected from biological selection pressures – and that it should move in the direction of greater knowledge, more complete sense, is taken too readily for granted. We cannot account for the 'unreasonable effectiveness' of mathematics in the world.[5] We are even further from being able to account for the 'unreasonable effectiveness' of our collective consciousness in making effective ordinary daily sense of the world. 'Out of the crooked timber of humanity, no straight thing was ever made' the German philosopher Immanuel Kant famously said.[6] Given the material out of which we are made and self-made, it is amazing just how straight we are at times and just how much of the world around us we seem to have got straight.

All this has been enacted in the theatre of a common world, or overlapping worlds, woven together out of headwinds and other communications between heads. By putting our heads together, we have been able to achieve what Munchausen only boasted of:

lifting himself up by his hair. Our heads have lifted themselves above the organic material of which they themselves are made. Humans have made themselves at home in organic bodies that could not have conceived of the things that fill the lives those bodies now permit. Humankind has increasingly made the world its own thing. Far from bowing our heads in shame, we should hold our heads up high.

Notes

Foreword

1. Ludwig Wittgenstein 'Preface' to *Philosophical Investigations*, translated by G. E. M. Anscombe (Oxford: Basil Blackwell, 1953).
2. Edmund Husserl quoted in David Bell, *Husserl* (London and New York: Routledge, 1990), p. 232.
3. Hippocrates, *On the Sacred Disease*, quoted in J. D. Spillane, *The Doctrine of the Nerves* (Oxford: Oxford University Press, 1981).
4. See, for example, Raymond Tallis *The Explicit Animal* (London: Macmillan, 1991, 1999); *Why the Mind is Not a Computer* (Exeter: Imprint, Academic, 2005); and *The Knowing Animal: A Philosophical Inquiry into Knowledge and Truth* (Edinburgh:Edinburgh University Press, 2005).
5. Raymond Tallis, 'Trying to find consciousness in the brain', *Brain*, 2004, vol. 127, pp. 2558–63.

1 Facing Up to the Head

1. This is remarked upon in Milan Kundera's discussion of the novel in *The Art of the Novel*, translated by Linda Asher (London: Faber and Faber, 1988).
2. Stéphane Mallarmé, from a letter written in his early twenties, quoted in Introduction to *Mallarmé*, introduced and translated by Anthony Hartley (London: Penguin, 1977), p. 14.
3. Evelyn Waugh, *Decline and Fall* (1928) (London: Penguin, 1937).
4. Most productively, that recently deceased philosophical genius Sir Peter Strawson, where it is discussed in his first major book, *Individuals. An Essay in Descriptive Metaphysics* (London: Methuen, 1959).
5. For the present, let us also note that anyone who scoffs at the notion of some kind of identity of the 'I' with my head has got a lot of arguments on his side. There are those who, with good reason, could dismiss our discussion as naïve and confused – confusing, for example, a supposedly

objective state of affairs – what I am – with some impressions I may have about what or where I am when I work myself up into a heightened state of introspection. They may have a point but the ideas discussed in that section are not uniquely naïve. They are just expressed rather nakedly. Neurophilosophers, who command great respect, and who find the self in the brain or in bits of it have two things on their side: they arrive at their conclusions by a route that is remote from introspection; and they wrap up their ideas in rather fancy jargon. The notion that the self (or the sense of the self) is located in some part of the brain – for example 'the cerebral cortex' – sounds more impressive than 'in my head' because it isn't a conclusion anyone could have arrived at unaided. It lies at the top of a massive pyramid of knowledge, obtained by means of sophisticated technologies and requiring the cooperative activity of vast numbers of individuals and institutions over many decades. The intuitions that drive the thinking behind neurophilosophy, however, are no less naïve, and considerably more vulnerable than the kind of ego-head talk we have been indulging in.

2   The Secreting Head

1. Philip Larkin, 'Ignorance' in *The Whitsun Weddings* (London: Faber and Faber, 1964).
2. I am grateful to the anonymous author of the excellent Wikipedia article (en.wikipedia.org/wiki/Cerumen) for some of the facts in these paragraphs.
3. Niall Ferguson, *The War of the World* (London: Allen Lane, 2006).
4. Quoted in Roy Porter, *Enlightenment: Britain and the Creation of the Modern World* (London: Allen Lane, 2000), p. 370.
5. Michael Steen, *The Lives and Times of the Great Composers* (Cambridge: Icon, 2003), p. 135.
6. See Raymond Tallis, *The Hand: A Philosophical Inquiry into Human Being* (Edinburgh: Edinburgh University Press, 2003).
7. Roland Barthes, *Mythologies*, selected and translated from the French by Annette Lavers (London: Jonathan Cape, 1972), p. 27. Thinking does not always demand sweat. One of the greatest philosophers of all time, the German Immanuel Kant, is said not to have sweated. This, for some, was of a piece with the abstraction of his thought, his rather mechanical routines, the passionless life he lived, his imperviousness to the charms of music, art, nature and the female sex. A sweatless man, he seemed to

have as little blood passing through his veins as in the logical subject, the transcendental ego, which, he proposed, synthesized the consciousness of human beings into a unity.

8. Jean-Paul Sartre, *Nausea*, translated by Robert Baldick (London: Penguin, 1965), p. 143.
9. Paul Broks, *Into the Silent Land: Travels in Neuropsychology* (London: Atlantic, 2003).
10. Norbert Elias, *The Civilizing Process: The history of manners and state formation and civilization* (Oxford: Blackwell Publishing, 1994). I came across this literature in Lynne Truss's wise and witty (and not at all grumpy) lamentation on the decline of manners in contemporary life, *Talk to the Hand: The Utter Bloody Rudeness of Everyday Life* (London: Profile, 2005).
11. G. F. Handel, *Messiah*, Part 2 Air (Alto).
12. George Steiner, *A Reader* (London: Penguin, 1984), p. 246.
13. Isaiah 50: 6.
14. Thomas Mann, *Tonio Kroger in Collected Stories*, translated by H. T. Lowe-Porter (London: Secker & Warburg, Everyman's Library, 2001), pp. 212–13.
15. The evidence thirty or so years later, after twenty-three clinical trials, is that it has moderate clinical benefits. (P. J. Poole and P. N. Black, 'Mucolytic agents for chronic bronchitis or chronic obstructive pulmonary disease', *The Cochrane Library*, Issue 2, 2006.)
16. Peter Godwin, *When a Crocodile Eats the Sun: a Memoir* (London: Picador, 2007), p. 165.
17. Alfred Lord Tennyson, 'Tears, idle tears, I know not what they mean' in *The Princess*, 1847.
18. I am indebted here to Silvia H. Cardoso and Renato M. E. Sabbatini, 'The Animal that Weeps', *Cerebrum*, 2000, 4(2), pp. 7–22.
19. Cardoso and Sabbatini (ibid.) quote Louis Bolk, a German anatomist, who describes humans 'as fetuses who are capable of reproduction'. No comment.
20. W. H. Frey and M. Langseth, *Crying: The Mystery of Tears* (Minneapolis: Winston Press, 1985).
21. Cardoso and Sabbatini, op. cit.
22. A most wonderfully witty and sympathetic account of this is to be found in Roy Porter's *Enlightenment: Britain and the Creation of the Modern World* (London: Penguin, 2000), pp. 281–94.

23. David Edmonds and John Eidinow, *Rousseau's Dog: A Tale of Two Great Thinkers at War in the Age of Enlightenment* (London: Faber and Faber, 2006), pp. 125–6.

First Explicitly Philosophical Digression

1. René Descartes, *Meditations on First Philosophy: Meditation VI, in The Philosophical Works of Descartes* (vol.1), translated by Elizabeth Haldane and G. R. T. Ross (Cambridge: Cambridge University Press, 1967), p. 192. Some eminent Descartes scholars emphasize his backtracking on a sharp separation between body and mind. For example, John Cottingham, interviewed in Andrew Pyle (ed.), *Key Philosophers in Conversation. The Cogito Interviews* (London: Routledge, 1999), p. 224.

> But if we want to understand what a human being is, as opposed to a kind of bloodless angel which just happened to be using a body (and this is a distinction Descartes often discusses), then we have to focus on bodily sensations and passions as key sources of evidence for the fact that we are not just minds inhabiting our bodies but are, as Descartes puts it, intimately united with them.

2. Ibid., p. 192.
3. Erich Heller, *The Disinherited Mind* (London: Penguin, 1961), p. 177.
4. Gilbert Ryle, *The Concept of Mind* (London: Penguin Books, 1963).
5. See Jean-Paul Sartre, *Being and Nothingness: An Essay on Phenomenological Ontology*, translated by Hazel Barnes (London: Methuen, 1957).
6. I can't seem to leave this topic. I want to record that, while explicit awareness is a necessary (if not a sufficient) condition of a bodily part being one's self, a curious distance opens up when one notices oneself indirectly – through mirrors, the expression on another's face, or words.

3 The Head Comes To

1. See Raymond Tallis, 'Complexity' in *Why the Mind is Not a Computer: A Critical Dictionary of Neuromythology* (Exeter: Imprint Academic, 2005)
2. See Armand LeRoi, *Mutants* (London: HarperCollins, 2003).
3. Quoted in Rüdiger Safranski, *Martin Heidegger: Between Good and Evil*, translated by Ewald Osers (Cambridge, Mass.: Harvard University Press, 1998).

4. 'The Savage Seventh' in Philip Larkin, *Required Writing: Miscellaneous Pieces 1955–1982* (London: Faber and Faber, 1983).

5. Gillian Butler and Freda McManus, *Psychology: A Very Short Introduction* (Oxford: Oxford University Press, 1998), p. 72.

6. See Daniel Povinelli, *Folk Physics for Apes* (Oxford: Oxford University Press, 2000).

7. Vladimir Nabokov, *Speak, Memory: An Autobiography Revisited* (London: Penguin, 1969), p. 17.

4 Airhead: Breathing and Its Variations

1. Dudley Young, *Origins of the Sacred: The Ecstasies of Love and War* (London: Little, Brown, 1992), p. xxi.

2. William Hazlitt, *Lectures on the English Comic Writers* (1818) (London: Oxford University Press, 1907).

3. It has been suggested that 'is' and 'ought' are located in separate cerebral hemispheres. V. S. Ramachandran, 'The evolutionary biology of self-deception, laughter etc', *Medical Hypotheses*, 1996, 47, pp. 346–62. By the time readers reach the end of this book, they will be well equipped to dispose of this and similar neurotheological claims.

4. Vic Gatrell, *City of Laughter: Sex and Satire in Eighteenth Century London* (London: Atlantic Books, 2006), p. 166.

5. Vladimir Nabokov, *Speak, Memory: An Autobiography Revisited* (London: Penguin, 1969), pp. 18–19.

6. Robert R. Provine, 'Laughter', *American Scientist*, Jan–Feb 1996, pp. 38–47.

7. Dutiful laughter – one of many examples of the human behaviour in which we simulate spontaneous delight – goes well beyond situations of subservience. I came across a reference recently to 'A sprinkle of mirth-less intellectual laughter, of the kind one hears at bookshop readings' in Zadie Smith's *On Beauty* (London: Penguin, 2006), p. 327. Many an author will smile with recognition. There is a marvellous discussion of 'dutiful laughter' (and much else relevant to our present theme) in Vic Gatrell's *City of Laughter* (see note 4). Gatrell (p. 159) quotes a story by Dr Johnson, describing a newcomer to London who has realized that the path to being accepted as a gentleman was to learn to laugh at the right time – not out of merriment but as 'one of the arts of adulation'. From 'laughing to shew that I was pleased, I now began to laugh when I wished to please'.

8. Provine, op. cit.

9. See chapter 10.

10. See, for example, S.-J. Blakemore, D. M. Wolpert and C. D. Frith, 'Central cancellation of self-produced tickle sensation', *Nature Neuroscience*, 1998, 1, pp. 635–40.

11. See Raymond Tallis, *The Knowing Animal: A Philosophical Inquiry into Knowledge and Truth* (Edinburgh: Edinburgh University Press, 2005).

12. Angus Trumble, *A Brief History of the Smile* (New York: Basic Books, reissued with a new Preface, 2004), pp. 93–4.

13. *As You Like It*, Act I, Scene II, ll. 52–3.

14. Camille Paglia, *Sexual Personae: Art and Decadence from Nefertiti to Emily Dickinson* (London: Yale University Press, 1990).

15. Rüdiger Safranski, *Martin Heidegger: Between Good and Evil*, translated by Ewald Osers (Cambridge, Mass.: Harvard University Press, 1988), p. 193.

16. V. M. Pomeroy, C. A. Clark, J. S. G. Miller, J-C Baron, H. S. Markus, R. C. Tallis, 'The potential for utilizing the "mirror neurone system" to enhance recovery of the severely affected upper limb early after stroke. A review and hypothesis', *Neurohabilitation and Neural Repair*, 2005, 19 (1), pp. 4–13.

17. See 'V Salutation after Sneezing' in 'The Magic of the Horseshoe', <http://www.sacred-texts.com/etc/hs/mhs47.htm>. Most of the lore that follows has been derived from this source.

18. Discussed by Leo Tolstoy in *War and Peace*.

5 Communicating With Air

1. 'The Physiology of Speech' in 'Speech', *Encyclopaedia Britannica*, 15th edn, vol. 28 (Chicago: University of Chicago, 1993), p. 79.

2. This figure, which is most often quoted, is bitterly disputed – not surprisingly, since language leaves no direct fossils. Some have pushed back the origin of speech to 100,000 years BP. There is little support for the claim that the word began around 1,000,000 BP.

3. Elsewhere I have suggested that it has arisen from the sense of the hand as a tool and that tools are the proto-linguistic link between animal grunts and human speech. How plausible that is, the reader may judge by consulting the books in which I have set out this hypothesis: *The Hand: A Philosophical Inquiry into Human Being* (Edinburgh: Edinburgh University Press, 2003), the first volume of the trilogy *Handkind*, in

which I reaffirm the huge gap between us and all other animals and try to explain the origin of that gap.

4. Robert Provine, 'Laughter', *American Scientist*, Jan–Feb 1996, pp. 38–47.

5. Gary Larson, the cartoonist, has a character, one Professor Schwarzkopf, who is the first human being to understand what dogs are really saying. Translating their speech, he finds they are all saying the same, single thing: 'Hey! Hey! Hey! Hey!' This story is recounted in Clive D. L. Wynne's balanced and sceptical *Do Animals Think?* (Princeton, NJ: Princeton University Press, 2004), p. 137.

6. Anne Karpf, *The Human Voice: The Story of a Remarkable Talent* (London: Bloomsbury, 2006), p .3.

7. Anon., quoted in Dorothy Robertson, 'Maintaining the art of conversation in Parkinson's disease', *Age and Ageing*, 2006, 35, p. 211.

6   Communicating Without Air

1. Georg Christoph Lichtenberg, *Aphorisms*, translated with an Introduction and Notes by R. J. Hollindale (London: Penguin, 1990).

2. M. H. Johnson and J. Morton, *Biology and Cognitive Development: The Case of Face Recognition* (Oxford: Blackwell, 1991), cited in H. Rudolph Schaffer, *Introducing Child Psychology* (Oxford: Blackwell, 2003), pp. 67–8.

3. See, for example, M. B. Lewis and A. J. Edmonds, 'Face detection: Mapping Human Performance' in *Perception* 2003, 32, pp. 903–20.

4. For more on this fascinating and tragic condition, see the National Institute for Neurological Diseases and Stroke website on 'Prosopagnosia'.

5. P. Ekman and G. Yamey, 'Emotions revealed: recognising facial expressions', *British Medical Journal*, 2004, 328, pp. 75–81.

6. Julien Guthrie, 'The lie detective: SF psychologist has made a science of reading facial expressions', *San Francisco Chronicle*, 16 September 2002.

7. Angus Trumble, *A Brief History of the Smile* (New York: Basic Books, with a new Preface, 2004), pp. xxxiii–xxxiv.

8. T. S. Eliot, 'The Love Song of J. Alfred Prufrock' in *Collected Poems 1909–1962* (London: Faber and Faber, 1963).

9. Trumble, op. cit., p. 73. The background to the perceived unseemliness of laughter and the superiority of the smile is described in fascinating detail in 'Laughing Politely' in Vic Gatrell's *City of Laughter: Sex and Satire in Eighteenth Century London* (London: Atlantic Books, 2006).

10. This insight into the tragic, fundamental project of humanity will live on.

11. Alexander Pope, 'An Epistle to Dr. Arbuthnot' (lines 315–16). Quoted in Trumble, op. cit., p. 133.
12. M. Katsikis and I. Pilowsky, 'A study of facial expression in Parkinson's disease using a novel micro-computer-based method', *Journal of Neurology, Neurosurgery and Psychiatry*, 1988, 51, pp. 362–6.

7 Notes on the Red-Cheeked Animal

1. J. K. Wilkin, 'Why is blushing limited to a mostly facial cutaneous distribution?', *Journal of the American Academy of Dermatology*, 1988, 19: 309–13.
2. Andrea Ladd, 'Ask a Scientist' Question Archives, <http://www.hhmi.org/cgi-bin/askascientist/highlight.pl?kw=blush&file=answers%2Fgeneral %2Fans_029.html>
3. Ibid.
4. As William Worthington pointed out, the invention of writing was the most wonderful of all human discoveries, making us 'Masters of other Men's labours and studies, as well as of our own', from 'An Essay on the Scheme and Conduct, Procedure and Extent of Man's Redemption', quoted in Roy Porter, *Enlightenment: Britain and the Creation of the Modern World* (London: Penguin, 2000), p. 74.
5. This notion, originating with the Pre-Socratic philosophers and most famously reiterated by Aristotle, that writing is a representation of speech – and hence a symbol of a symbol – is not entirely uncontested. See, for example, Roy Harris, *The Origin of Writing* (London: Duckworth, 1986).
6. As a northerner writing this sentence in the city of Bath, I am particularly conscious of this at present.

8 The Watchtower

1. Much of the discussion in this chapter – particularly on the nature of human consciousness and knowledge – recapitulates arguments and ideas developed at length in my trilogy Handkind, in particular the third volume, *The Knowing Animal. A Philosophical Inquiry into Knowledge and Truth* (Edinburgh: Edinburgh University Press, 2005).
2. Gordon Bowker, *Inside George Orwell* (London: Palgrave Macmillan, 2003).

3. S. Hecht, S. Shlaer, M. H. Pirenne, 'Energy, Quanta and Vision', *Journal of General Physiology*, 1942, 25, pp. 819–40.

4. For a fascinating critical review of the recent literature, see Norma Graham 'Light Adaptation', <http://www.columbia.edu>

5. See, in particular, 'An Outline of Epistogony', chapter 4 of *The Knowing Animal*, op. cit.

6. Of course, when I move my head, things may seem to move because they change their relationships to one another. As I tilt my head, the objects concealed by my laptop screen change. Instead of the notice on the wall, it is the telephone to the right of it that is concealed. I know, however, that this is not actual movement, only a change in the appearance of the relationships between the objects, the way those relationships are presented to me.

7. P. F. Strawson, *Individuals. An Essay in Descriptive Metaphysics* (London: Methuen, 1964), p. 65.

8. Incidentally, the fact that knowledge, which is based on the idea that there are objects, always exceeds its basis in our sense experiences is not a scandal putting either knowledge or our senses in question; rather it is of the very nature of knowledge.

9. The case for the key role of the hand in driving the journey of human consciousness from animal sentience to human knowledge, and the awakening of the human organism to its own body so that it apprehends it and becomes an embodied subject, is set out in the first volume of the trilogy referred to in note 1: *The Hand. A Philosophical Inquiry into Human Being* (Edinburgh: Edinburgh University Press, 2003).

10. The mystery by which the gaze looks out – one that cannot be solved by the input of light (or 'light energy') – is highlighted in 'The Troubles with Neurophilosophy', section 2.2 of *The Knowing Animal*, op. cit.

11. W. D. Ross (ed.), *Aristotelis De Anima* (Oxford: Clarendon Press, 1956), 424a, pp. 17–19.

12. Readers wishing to pursue this train of thought further may like to read 'The Difficulty of Arrival' in my *Newton's Sleep. Two Cultures and Two Kingdoms* (London: Macmillan, 1995).

13. Letter, 16 December 1911, quoted in Erich Heller, *The Disinherited Mind. Essays in Modern German Literature and Thought* (London: Penguin Books, 1961), p. 175.

14. Rüdiger Safranski, *Martin Heidegger: Between Good and Evil*, translated by Ewald Osers (Cambridge MA: Harvard University Press, 1998), p. 103.

15. These ideas are further developed in 'The Work of Art in an Age of Electronic Communication' in Raymond Tallis, *Theorrhoea and After* (London: Macmillan, 1999).

16. Martin Heidegger, *Letter on Humanism in Basic Writings*, edited by David Farrell Krell (London: Routledge & Kegan Paul, 1978), p. 237.

17. See Jean-Paul Sartre, 'The Look' in *Being and Nothingness: An Essay on Phenomenological Ontology*, translated by Hazel Barnes (London: Methuen, 1957), pp. 252–302.

18. This is discussed in 'Deindexicalised Awareness' in *The Knowing Animal*, op. cit., especially pp. 119–122.

19. It is interesting, but unsurprising, that such individuals do not point things out to others either. Pointing, which is one of the most important and fundamental modes of communication with others, antedates speech in normal children by a few months.

20. Leo Tolstoy, *War and Peace*, translated by Anthony Briggs (London: Penguin, 2005), p. 1069.

21. John Gay, libretto for G. F. Handel, *Acis and Galatea*, A Masque in Two Acts, c. 1718.

22. Dante Alighieri, translated by George R. Kay in *The Penguin Book of Italian Verse* (London: Penguin, 1958), p. 85.

23. Leo Tolstoy, *War and Peace*, op. cit., p. 622.

24. John Richardson, *A Life of Picasso: Volume 1, 1881–1906* (London: Jonathan Cape, 1991).

25. David Gilmore, *Aggression and Community. Paradoxes of Andalusian Culture* (New Haven: Yale University Press, 1987).

26. I owe the phrase to Thomas Nagel, whose *The View from Nowhere* (Oxford: Oxford University Press, 1986) argued that the ultimate aim of the scientific endeavour to arrive at impersonal or objective truth was to reach 'a view from nowhere'.

27. Chris Frith, *Making up the Mind: How the Brain Creates Our Mental World* (Oxford: Blackwell Publishing, 2007), p. 143.

9 The Sensory Room

1. I owe this phrase to John Searle, *Mind: A Brief Introduction* (Oxford: Oxford University Press, 2004).

2. I am grateful to a delightful little line drawing in *Encyclopaedia Britannica*, 15th edn, vol. 27, p. 206.

3. Stephen B. Karch, quoted in Anthony Daniels, *Romancing Opiates. Pharmacological Lies and the Addiction Bureaucracy* (New York: Encounter Books, 2006), p. 22.
4. Franz Kafka, *The Diaries of Franz Kafka, 1910–1923*, edited by Max Brod, translated by Joseph Kresh and Martin Greenberg (London: Penguin, 1964), p. 10.
5. I am very indebted to Tim Jacob for his excellent 'Taste – A Brief Tutorial', <http://www.cf.ac.uk.biosi/staff/jacob/teaching/sensory/taste.html> upon which I have drawn for much of what follows.
6. Ibid.
7. T. S. Eliot, 'Little Gidding' in *Collected Poems 1909–1962* (London: Faber and Faber, 1963), p. 218.
8. I owe this fascinating sidelight on taste to Michael Berry, op. cit.
9. Richard E. Cytowic, *The Man Who Tasted Shapes* (Cambridge, Mass.: MIT Press, 1998).
10. S. J. Blakemore, D. Bristow, G. Bird, C. Frith, J. Ward, *Brain*, 2005, 128, pp. 1571–83.
11. The writers belonging to the Symbolist movement that reached its height towards the end of the nineteenth century were obsessed with synaesthesia. Arthur Rimbaud's 'Vowels' connected each of the vowels with a colour. Des Esseintes, the hero of J-K Huysmans' *Against Nature* dreamt of a 'mouth organ' that would dispense liqueurs, each corresponding to the sound of a particular orchestral instrument. He would use it to compose gustatory symphonies. Arthur Symons praised the superior sensibility of the Opium Smoker who had the ability to sense

    > Soft music like a perfume, and sweet light
    > Golden with audible odours exquisite.

12. There are other explanations of the especial mnestic potency of smell. The most favoured is that the pathways mediating smell in the brain pass into the evolutionarily most primitive and ancient parts of the brain – the hypothalamus, the hippocampus and the brainstem. I think this misses the point: the isolated brain cannot account for the world that we recognize as our own.
13. Marcel Proust, *Remembrance of Things Past*, translated by C. K. Scott Moncrieff and Terence Kilmartin (London: Penguin, 1983), pp. 50–51.
14. This, and some of the other facts that follow, are taken from Tim Jacobs, 'Smell (Olfaction). A Tutorial on the Sense of Smell', Cardiff University.

15. Press release for the Nobel Prize for Physiology or Medicine, 4 October 2004.
16. K. Stern and M. McClintock, 'Regulation of ovulation by human pheromones', Nature, 1998, 392, pp. 177–9.
17. Beverley Strassmann, 'Menstrual synchrony pheromones: cause for doubt', Human Reproduction, 199, 14(3), pp. 579–80.
18. Quoted in Michael Steen, The Lives and Times of the Great Composers (Cambridge: Icon Books, 2003), p. 702.

Third Explicitly Philosophical Digression

1. The opacity of the head was the basis of a joke by a comedian whose name I cannot recall. He complained about breast-feeding in public on the grounds that 'the baby's head gets in the way of the view'.
2. The case for this claim is developed at length in Raymond Tallis, The Hand. A Philosophical Inquiry into Human Being (Edinburgh: Edinburgh University Press, 2003).
3. In Paul Valéry's Monsieur Teste, the eponymous hero thinks to himself: 'My weight, what a possessive!'. Paul Valéry, Monsieur Teste, translated by Jackson Mathews (London: Routledge & Kegan Paul, 1973), p. 49.
4. Gabriel Marcel, 'Outlines of a Phenomenology of Having' in Being and Having (London and Glasgow: Fontana, 1965), p. 179. The translation is not attributed, which is a serious injustice because it is excellent.
5. Ibid., p. 169.
6. Hazel E. Barnes, 'Sartre's ontology: The revealing and making of being' in The Cambridge Companion to Sartre, edited by Christina Howells (Cambridge: Cambridge University Press, 1992), p. 21.

10   Head Traffic

1. Richard Dawkins in Key Philosophers in Conversation. The Cogito Interviews, edited by Andrew Pyle (London and New York: Routledge, 1991).
2. W. H. Auden, 'Tonight at Seven-Thirty', section X in 'Thanksgiving for a Habitat', in Selected Poems (London: Faber and Faber, 1968), p. 136.
3. I owe this, and some other observations in the succeeding paragraphs, to Robin Lane Fox, 'Food and Eating: An Anthropological Perspective', Social Issues Research Centre, Vox Rationis, 11 April 2007.

4. Karl Marx and Frederick Engels, *The German Ideology*, edited with an Introduction by C. J. Arthur (London: Lawrence & Wishart, 1974), p.42.
5. Surprising but true. Caroline Wyatt, 'Mastering French Manners the Hard Way', *From Our Own Correspondent*, BBC4, Saturday, 23 December 2006.
6. Aletheia Jackson, 'For the love of Whizdom', *Critical Review*, 1988, vol. 4 (3). It is a review of Robert Nozick's *The Examined Life. Philosophical Meditations*.
7. I have been greatly helped in this section by R. Bowen 'The Physiology of Vomiting', from <http://www.vivo.colostate.edu>.
8. See Raymond Tallis, 'Communication, Time, Waiting' in *Hippocratic Oaths. Medicine and its Discontents* (London: Atlantic Books, 2004).
9. Anonymous editorial, 'Psychogenic Vomiting', *British Medical Journal*, 12 August 1967, 3(5562), pp. 386–7.
10. Bob Newhart, 'Introducing Tobacco to Civilization', 1962.
11. Allan M. Brandt, *The Cigarette Century. The Rise and Fall, and Deadly Persistence of the Product that Defined America* (New York: Basic Books, 2007).
12. In this regard, it is interesting to note that smoking accounts for more than half of the difference in life expectancy and healthy life expectancy between rich and poor in the UK. (Data cited in *The Health of the Population*: 'Annual Report of the Directors of Public Health for Cornwall and the Isles of Scilly', 2006.)

11   Head on Head: Notes on Kissing

1. This paper and its boundless implications are discussed in James Gleick's wonderful *Chaos. Making A New Science* (London: Cardinal, 1988), see especially pp. 94 et seq.
2. Readers may wish to do their own calculation. They may not come to the same figure but they will arrive at essentially the same conclusion.
3. Sigmund Freud (1912), ''On the Universal Tendency to Debasement in the Sphere of Love' in *Five Lectures on Psycho-Analysis, The Standard Edition of the Complete Psychological Works of Sigmund Freud* (London: The Hogarth Press, 1957).
4. Richard Dawkins in Andrew Pyle (ed), *Key Philosophers in Conversation: The Cogito Interviews* (London and New York: Routledge, 1999).
5. Christopher Hamilton, *Living Philosophy. Reflections on Life, Meaning and Morality* (Edinburgh: Edinburgh University Press, 2001), p. 136.

6. William Ian Miller, *The Anatomy of Disgust* (Cambridge, Mass.: Harvard University Press, 1997), p. 137.
7. Paul Valéry, *Monsieur Teste*, translated by Jackson Mathews (London: Routledge & Kegan Paul, 1973), p. 49.
8. W. H. Auden from 'I Am Not a Camera', quoted in *Selected Poems* (London: Faber & Faber, 1971).

12 Headgear

1. I owe these lists to the essay on 'Mustache' in Victoria Sherrow's delightful *Encyclopaedia of Hair. A Cultural History* (Westport, Connecticut, London: Greenwood Press, 2006).
2. Rüdiger Safranski, *Martin Heidegger: Between Good and Evil*, translated by Ewald Osers (Cambridge, Mass. and London: Harvard University Press, 1998), p. 103.
3. There is an account of the so-called Alder Hey scandal in Raymond Tallis, *Hippocratic Oaths* (London: Atlantic, 2004).
4. G. G. Gallup, 'Self-recognition in primates: A comparative approach to the bidirectional properties of consciousness', *American Psychologist*, 1977, 32, pp. 329–38.

Fourth Explicitly Philosophical Digression

1. See Francis Ponge, *Soap*, translated by Lane Dunlop (Stanford, Calif.: Stanford University Press, 1988). Activities such as the bored purchase of soap-on-a-rope as a Christmas present for an unfavourite aunt should not be allowed to conceal from us the true miracle of this commodity, unstickying the world.
2. For the full story, see 'Hand talking to Hand' in Raymond Tallis, *My Hand: A Philosophical Inquiry into Human Being* (Edinburgh: Edinburgh University Press, 2003).
3. Raymond Tallis, *I Am: A Philosophical Inquiry into First-Person Being* (Edinburgh: Edinburgh University Press, 2004).

13 In the Wars

1. Niall Ferguson, *The War of the World* (London: Allen Lane, 2006), p. xxxvii.
2. 'An adult human cadaver head cut off around vertebra C3 with no hair weighs between 4.5 and 5 kilograms or about 8% of the whole body

weight', according to Danny Lee, an anthropologist. Add the weight of a gold plate and one realizes that Salome must have been a strong girl.

3. Bernard Porter, 'Thrilled to bits' review of Saul David's *Victoria's Wars: The rise of Empire* (London: Viking, 2006), *Times Literary Supplement*, 21 July 2006, p. 24.

4. William Shakespeare, *As You Like It*, Act II, scene I, ll.10–11.

5. There are hundreds of thousands of references to this famous case. The interested reader might like to consult A. T. Steegmann, 'Dr. Harlow's famous case: the "impossible" accident of Phineas P. Gage', *Surgery*, 1962, 52, pp. 952–8. The psychologist, Dr Stuart Derbyshire (personal communication) has pointed out to me that Gage may have been misrepresented by Harlow. After all, Gage managed to hold down a difficult job, and, in the absence of any record of a single act that he should have been ashamed of, one should perhaps respect the dignified way he dealt with an appalling disfigurement he had following his injury. This only underlines how there is more to human nature than brain science can capture.

6. Richard Clark, 'Beheading', <http://www.richard.clark32.btinternet.co.uk/behead.html>

14   The Dwindles

1. Edward Thomas, 'Lights Out', *Collected Poems* (London: Faber and Faber, 2004), ll. 1–6.

2. The fact that it is physiological and that we share it with animals makes it more, not less, mysterious. What adaptive purpose could it serve? By what means could a state of total invigilance and compulsory helplessness aid survival? The suggestion that sleep is adaptive because it immobilizes us in the hours of darkness and thereby makes us less visible, audible etc to predators is at odds with the observation that predators sleep more soundly than their prey: those who sleep safely sleep longer. The more widely accepted link between sleep and anabolic activity – the building up, maintenance and restoration of tissues – is hardly more compelling. There is absolutely no reason why loss of awareness should be essential for this – if only because awareness is not readily connected with the physical characteristics of the body. Anyway, although we share sleepiness with beasts, our sleep (as we shall discuss) is as different from the sleep of beasts by the same token as our reflexive consciousness is different from the first order consciousness of beasts.

3. This is the basis of the unfounded feeling that our body is, in itself, a place of darkness. Of course, *qua* physical object, it is neither light nor dark. For a further discussion of this, see Raymond Tallis, '(That) I am this (thing)' in *On the Edge of Certainty* (London: Macmillan, 1999).
4. Not so surprising, of course, because the eyelids are self-raising: the object they have to lift has the same weight as the object that is lifting them.
5. The German original reads thus:

> *Rose, oh reiner Widerspruch, Lust*
> *Niemandes Schlaf zu sein unter soviel*
> *Lidern.*

6. Hence, 'Bawdyhouse. I seek the kips where Shady Mary is' in James Joyce, *Ulysses*, new edn (London: The Bodley Head, 1960).
7. It is amusing to think that there is an entire system in the brain – the reticulo-cortical loop – devoted to sustaining wakefulness. One would like to think that this is a reflection of the humility of the creator in recognizing the hypnotic tedium of the fare he serves up to us in our senses and daily life!
8. In his profound essay 'The Need to Sleep' in *Living Philosophy: Reflections on Life, Meaning and Morality* (Edinburgh: Edinburgh University Press, 2001), Christopher Hamilton observes the fundamental difference between animal and human sleep:

> What we call sleep in an animal actually marks nothing more than a change in *degree* in the animal's consciousness, whereas in a human being sleep marks a change in *kind* from wakening consciousness. For animals never feel the need to be rid of themselves as a person does; they are never burdened by themselves. (pp. 148–9).

Animal sleep is an intermission in consciousness; while human sleep is an intermission in both consciousness and self- consciousness. Hamilton comments upon the remarkable lack of serious philosophical work on sleep (apart from some discussion of dreams in the philosophy of mind and epistemology) – which, after all, occupies a third of the lives of most people.

9. Ibid. Hamilton has much that is illuminating to say about insomnia.
10. Edward Thomas, 'Lights Out' in *Collected Poems*, op. cit. ll. 25–30.

11. Rebecca C. C. Brooke and Christopher E. M. Griffiths, 'Ageing of the Skin' in Raymond C. Tallis and Howard Fillit (eds), *Brocklehurst's Textbook of Geriatric Medicine and Gerontology* (6th edn, Edinburgh: Churchill Livingstone, 2003), p. 1269.

12. Thomas Hardy, 'I look into my glass', *Wessex Poems and Other Verses in Selected Shorter Poems*, chosen and introduced by John Wain (London: Macmillan, 1966).

13. When George Melly asked Mick Jagger how his face got so many creases, he said 'laughter lines'. 'Nothing is that funny,' Melly said.

14. William Shakespeare, *Anthony and Cleopatra*, Act I, scene 5, ll. 27–9.

15. Philip Larkin, 'Skin' in *Collected Poems* (London: Faber and Faber, 1988).

16. William Shakespeare, *Hamlet*, Act V, scene 1, ll. 206–9.

17. I am very grateful to the excellent – if somewhat stomach-churning – article 'Forensic Entomology: The Use of Insects in Death Investigations' by Dr Gail S. Anderson, <http://www.rcmp-learning.org/docs/ecdd0030.htm>.

18. Ibid.

19. Philip Larkin, 'Aubade' in *Collected Poems*, op. cit.

20. W. B. Yeats, 'Death' in *The Winding Stair and Other Poems* (London: Macmillan, 1933).

21. Rainer Maria Rilke(1928), *Rodin* (New York: Dover, 2006).

22. Quoted in Jonathan Barnes, *The Presocratic Philosophers* (London: Routledge, revised edition, 1982).

23. Lord Byron, *Don Juan*, Canto 1, Stanza 217.

24. Colette Sirat, *Writing as Handwork. A History of Handwriting in Mediterranean and Western Culture* (Belgium: Brepols-Turnhout, 2006).

25. Imre Kertész, 'Someone Else. A Chronicle of Change', translated by Tim Wilkinson, *Common Knowledge*, Spring 2004, 10, 2, pp. 314–46.

15 Head and World

1. Rainer Maria Rilke, *Duino Elegies*, 'The Eighth Elegy' in *Selected Works*, vol. II, translated by J. B. Leishman (London: The Hogarth Press, 1967), p. 242.

2. Ludwig Wittgenstein, *Tractatus Logico-Philosophicus*, translated by D. F. Pears and B. F. McGuinness (London: Routledge & Kegan Paul, 1961), p. 7.

3. Walt Whitman 'Song of Myself' in *Leaves of Grass* (1855).

4. See Raymond Tallis, *Why the Mind is Not a Computer* (Exeter: Imprint Academic, 2004).

5. Thomas De Quincey, 'The Pleasures of Opium' in *Confessions of an English Opium-Eater and Other Writings* (London: New English Library, 1966), p. 61.

6. Leo Tolstoy, *War and Peace*, translated by Anthony Briggs (London: Penguin Classics, 2006), p. 663.

7. Leszek Kolakowski, 'Conscience and Social Progress' in *Marxism and Beyond: On Historical Understanding and Individual Responsibility*, translated by Jane Zielonko Peel (London: Paladin, 1971), p. 156.

16   The Thinking Head

1. Anyone interested in the Thinking Head should read at least a few pages of *Ulysses* – the greatest novel of the twentieth century.

2. Or, as the Surrealists used to say, '*La pensée se forme dans la bouche*' – the thought is formed in the mouth.

3. This discussion has not yet clearly differentiated between thought-types, such as the general thought that there is a cat in the room next door and token thoughts corresponding to the general thought when it is had on particular occasions. This is an important omission and will be only partly remedied in the next section. Anyone who wants to look into these matters further may wish to consult Raymond Tallis, *I Am. A Philosophical Inquiry into Human Being*, passim.

4. The arguments are beautifully summarized in Nicholas Fearn, *Philosophy. The Latest Answers to the Oldest Questions* (London: Atlantic, 2005).

5. The *locus classicus* of this approach is Gilbert Ryle's *The Concept of Mind* (London: Peregrine, 1963) which still retains its power to exasperate nearly sixty years after it was first published.

6. This argument is borrowed from Quassim Cassam's marvellous *Self and World* (Oxford: Oxford University Press, 1997). See especially 'The Objectivity Argument', section 9, 'Core-Self and Bodily Self'.

Epilogue

1. For a brilliant exposition and critique of these trends, read Kenan Malik, *Man, Beast and Zombie. What Science Can and Cannot Tell Us About Human Nature* (London: Weidenfeld and Nicolson, 2000).

2. Niall Ferguson, *The War of the World: History's Age of Hatred* (Penguin: London, 2006).

3. See John Gray, *Straw Dogs: Thoughts on Humans and Other Animals* (London: Granta: London, 2002).

4. For an excellent account of the disintegrated selves of people with autism, see Charlotte Moore, *George and Sam: Autism in the Family* (London: Penguin, 2005).

5. Eugene Wigener, 'The Unreasonable Effectiveness of Mathematics in the Natural Sciences' in *Communications in Pure and Applied Mathematics*, vol. 13(1), February 1960.

6. Immanuel Kant, quoted in the epigraph to Isaiah Berlin, *The Crooked Timber of Humanity. Chapters in the History of Ideas*, edited by Henry Hardy (London: HarperCollins, Fontana, 1991).

Index

male 108, 151–2
mental 4
of others 43, 104, 145–55, 175, 212
as signal 155
Glasgow kiss 226, 228
Godwin, Peter 35–6
Goethe, J.W. von xx, 116
Goldsmith, Oliver 260
Grabbe, Christian Dietrich 54
grammar 94
Gray, John 289
grooming 222–5
growth, cognitive 57–60
grunting 87–8
gustducin 173

hair 212–18
    awareness of 9–10
    greying 245
    growth 4
    and identity 217–18
    philosophy of 214–17
Hamilton, Christopher 208, 307 n.8
Han, Ran, Benaroia, Mark and Goldlist, Barry 236
Handel, G. F., *Messiah* 30–1
Hardy, Thomas 246, 307 n.12
harrumphing 87–8
hawking 30
Hazlitt, William 68
head
    construction 51–7
    damage to 230–5
    instrumental use 180–2, 184
    knowledge of 7–8, 257–9
    as in other heads 270–3
    protection 229–30

as weapon 226–9
in the world 260–4
as in the world 260–4
world as in 264–70
head butt 226–8
heading 228
hearing 11, 129, 136, 161–70
    and head sounds 177–9
    mechanisms 162–5, 229
    as mystery 167–9
    and perception of sound 165
    of speech 165–7
    of thoughts 4, 265, 276–7
Hegel, G. W. F. 145
Heidegger, Martin xv, 13, 82, 143, 145, 217
helmets 229–30
Hemingway, Ernest 199
Heraclitus 244
hiccup 88
hippocampus 302–3 n.12
Hippocrates xviii
Hofmann, Michael 109
Housman, A. E. 51, 65, 236
humanism xiv
humanity
    achievements 289–91
    as beasts or zombies 288–9
hunger 100
Husserl, Edmund xv–xvi, 12
Huysmans, J.-K. 302 n.11
hypothalamus 21, 23–4, 172, 302–3 n.12

identity
    and hair 217–18
    and identification 104–5
immortality 251–6

snot/snotty 34–5, 71
sobbing 37, 38, 41
Sorites paradox 214, 215
soul 46, 67, 289
sound *see* hearing
space
    egocentric 12–14, 143
    hodological 261
    location in 260–4, 265
    and thought 279–80
    and vision 135–6, 140
Space of Possibility 56, 60, 89, 90, 96–7, 283
spectacles 220
shades 43–4, 153
speech 90–8, 120–2
    and breathing 68
    and common cold 33–4
    and hearing 165–7
    and laughter 94–5
    origins 93–4, 120
    reported 121–2
    and writing 97, 122–3
    see also thought
Speke, John Hanning 85
spelling 124–5
spirit
    and breath 67
    and speech 91
spit, spitting 28–32, 227
spluttering 88
sputum
    and mucus 30, 34–5, 197
    and saliva 27, 29–31
stapes (stirrup) 162
staring 112
Steen, Michael 22–3
Steiner, George 31

stereocilia, auricular 164–5
Storr, Anthony 27–8
Strassmann, Beverley 177
Strawson, Peter 136, 293 n.4b
subjectivity, and philosophy xvi
suffering 97–101
sweat 19–24
    stress-related sweating 23–4
    thermal sweating 19–23, 24
Symons, Arthur 302 n.11
synaesthesia 174

taste 11, 170–2
    and smell 174, 175
    and synaesthesia 174
taste buds 171–3
tears 4
    as emotional reaction 37–44, 226, 259, 283–4
    as physical defence 36–7
    and sneezing 85
teeth
    brushing 223–5
    grinding 40
    growth 4
    as weapon 227
telereceptors 11, 14
Tennyson, Alfred, Lord 36, 39, 202, 295 n.17
thirst 100
Thomas, Edward 238, 245, 306 n.10
thought 4–7, 274–87
    as audible 4, 265, 276–7
    awareness of 7
    and brain 280–2
    and breathing 67–8, 198
    control 285
    location 4–5, 275–6, 277–82

as mobile 285–6
and philosophy 286
and self 5, 46–50
and sweating 23–4
and words 92, 97, 122–4, 275, 278, 284
see also speech
tickling 68, 72, 73–4
time
and ageing 245–51
and eating 188–9
and eternity 251–2
and expectation 69–70, 202–5, 210
tinnitus 9, 178–9
tiredness 100
Tolstoy, Leo 42
War and Peace 150–2, 270
tongue 170–1
piercing 220
touch 11, 135, 136, 137
tranquillizers 233–4
transducin 173
Trumble, Angus 110
Truss, Lynn 295 n.10
tut-tutting 88
Twain, Mark 116

umami 171
utricle 162–3

Valéry, Paul 1, 183, 209, 259
viewpoint, and self 12, 135, 138, 142–3, 148, 161, 283
vision 11–12, 129–55
biology 131–5
and cognition 137–43
and distance 135–6

and egocentric space 12–13
and information-gathering 129–30
and language 140
see also gaze
vitreous humour 132–3
voice 96, *see also* speech
vomiting 4, 193–9
and laughter 195–6
psychogenic 195

waking 60–3, 242
washing the face 222–5
Waugh, Evelyn 3
weapon, head as 226–9
Weil, Simone 41
Wells, H. G. 245
whistling 88
Whitehead, Alfred North 43
Whitman, Walt 265
Who Wants to be a Millionaire? 81
Wilde, Oscar 29
Williams, Emlyn 41
winking 107–8, 117
Wittgenstein, Ludwig xiv–xv, 105, 145, 170, 263
Worthington, William 299 n.4b
wrinkles 245–7
writing
and alphabetization 123–6
and speech 97, 122–3
and spelling 124–5

Xenophon, *Anabasis* 85–6

yawning 81–4
as infectious 81, 82–3
and laughter 81–2